DEAD HEAT

DEAD HEAT
Global Justice and Global Warming

TOM ATHANASIOU
and PAUL BAER

AN OPEN MEDIA BOOK

SEVEN STORIES PRESS / NEW YORK

363,738
ATH

A Seven Stories Press First Edition. An Open Media Book.

Series Editor, Greg Ruggiero.

ISBN: 1-58322-477-7

COVER DESIGN: Greg Ruggiero
COVER PHOTO: AP Photo/Dave Martin

9 8 7 6 5 4 3 2 1

Printed in Canada.

To Anil Argawal, an honest man.

ACKNOWLEDGMENTS

Thanks to Steve Bernow, Surje Desai, John Gershman, Bill Hare, Barbara Haya, Patrick McCully, and Oras Tynkkynen for reading early versions of some chapters. Thanks to Greg Ruggiero at Seven Stories for encouraging us to write this book, and to the supporters and board members of EcoEquity for supporting the collaboration that made it possible. And thanks to countless members of the Climate Action Network, for everything.

CONTENTS

OUR PROBLEM, AND YOURS

Setting out to write *Dead Heat*, we had a problem.

In this book, we argue that the battle against global warming is key to the larger battle for global justice; that its outcome may, in fact, be almost as decisive politically as it will be ecologically. We argue, moreover, that this is true for a very particular reason: because there can be no workable climate-protection regime without a *historic compromise* between the rich world and the poor, one that actually specifies the terms by which we will share the Earth's very limited "atmospheric space."

There's been a lot of nonsense written about ecological limits, so let's be clear here: the need to protect the atmosphere, and all the human and non-human beings that depend on it, demands that we establish an enforceable limit to the global emissions of greenhouse gases, and that we find *equitable ways* to share this limited space. So while we eagerly admit that this means the political defeat of the current U.S. administration's policy, such a defeat would hardly be sufficient. What we really need is a vast new wave of global cooperation, one that includes both the countries

9

that became wealthy through fossil-fuel-powered industrialization and the "developing counties" for whom such a path is now foreclosed. And we need this new wave of cooperation to emerge soon, even as the world sinks into frightening economic malaise, even as war threatens, even as the United States—the last superpower—does everything possible to block the emergence of global social and environmental controls.

Tough, that one. But it's not the problem.

The problem is that *we're already in very serious trouble.* Thus, our first task, as the authors of this book, must necessarily be the unhappy one of explaining just how bad the situation really is. Here then, right up front, is our punch line: The science shows, in mercilessly numeric terms, *that even if we move quickly to cap the emission of greenhouse pollutants, the consequences of global warming will soon become quite severe, and even murderous, particularly for the poor and the vulnerable. And in the more likely case where we move slowly, the impacts will verge on the catastrophic.*

That, briefly, is our problem. And yours.

Now we know that by saying this so clearly we open ourselves to dismissal by the climate "skeptics." But the skeptics can go to hell, and we're basically going to ignore them. For our goal here is exactly that which climate skepticism is designed to prevent: We aim to take a cold-eyed look at our

Tom Athanasiou and Paul Baer

actual predicament, and then, to the best of our abilities, to draw realistic conclusions about what must be done if the damage is to be kept within manageable limits.

In that sprit, *Dead Heat* argues that justice—not small change and rhetoric but real developmental justice for the people of the South—is going to be necessary, and surprisingly soon. It argues that we must, above all else, find a path to a *just sustainability*—one that works for the weak and the vulnerable as well as the rich and the strong—and that the essential problems of this path are global, or, rather, rooted in localities the world around. It argues, particularly, that we must make *a phased transition from the Kyoto Protocol (which, to be clear, we emphatically support) to a second-generation climate treaty based on equal rights to the atmosphere*. And it concludes that the fight for such a "rights-based" climate treaty will be a key event in the emerging battle for common global resources of all kinds, from the oceans to the freshwaters to the genome.

Dead Heat argues that without a global campaign for a rights-based climate treaty—one in which atmospheric overusers have to *pay* for their overuse of the atmospheric space—we're simply not going to be able to move fast enough. That we will, instead, dither and debate and lose ourselves in small wars of position, but that even as we do, the "party of the Fortress World" (think Bush, and see chapter 9) will gain in power, and the climate system will lock into a trajectory that makes a truly intolerable degree of warming almost inevitable. That to do better, we must see that *climate justice* and *climate realism* are two sides of one coin, and that only the two, together, can make the "soft-landing scenario" into something more than a crazy dream.

We'll be talking a lot about "soft landing" scenarios and "corridors" and "paths," so, again, we need to be clear. "Soft landing" is just a neat bit of climate movement jargon for a precautionary global transition in which we avoid catastrophic climate change. It implies specific limits on greenhouse-gas emissions (described in chapters 2 and 3) as well as a global turn towards justice and that still-imagined land we call "sustainability" and, of course, away from unilateralism and war. It implies, in other words, that there isn't going to be a soft *climate* landing that isn't also a soft landing in a whole lot of other dimensions as well.

Our premise, though, is that global warming is a pivotal problem, and that its solution, if we can contrive one, will be decisive. To make this argument effectively, we must be able to say useful and specific things about the way forward. If, after reading *Dead Heat*, you conclude that we've failed to do so, grant us at least that the task is not an easy one. Justice or not, any campaign for a new global climate regime must drive the rapid "decarbonization" of the economy, delivering fossil-fuel free development in the South and a renewable energy transition in the North, and it must do so

Tom Athanasiou and Paul Baer

within today's troubled, and emphatically capitalist, social and economic world.

It isn't going to be easy.

Look at it this way: Humans have increased the global average surface temperature by only 0.6 degrees Centigrade, but already the climate is changing fast. And, as we will show, *any future in which we manage to hold the warming to a maximum of 2°C (and 2°C would almost certainly mean massive suffering and destruction) will require decisive global action*, something like a global Marshall Plan, but tuned, particularly, to sustainable energy development.

And it will require it soon.

Dead Heat, then, is an argument, not an overview. In it we're going to make the case for climate equity, as we understand it. And we're going to argue that equity, in addition to all its manifold moral and humanitarian attractions, must be seen as the most "realistic" of virtues. We're going to argue, in fact, that equity—or justice if you prefer—is essential if we're going to get out of here in time.

Tom Athanasiou and Paul Baer
Albany, California
September 2002

CHAPTER 1

AN INTRODUCTION

We all know about global warming. Most of us even know that the world's emissions of greenhouse pollutants are the biggest problem. But while we curse the SUVs and their drivers, we rarely face the larger situation square on. What, after all, are we supposed to do about the overall logic of development, or about the rich-poor divide that ultimately defines it? What are we supposed to do about the fact that everyone wants dignity and prosperity, and that, these days, prosperity means ever-increasing gross domestic product (GDP), energy consumption, and greenhouse pollution?

And that rising greenhouse pollution means disaster?

It's a grim situation, and added to all the other troubles in the world, it doesn't encourage optimism. So, what to do? We can, of course, fall back on the traditional hope of capitalist man—technological redemption—and doing so is not entirely mad. Despite the best efforts of the fossil fuel cartel and the Bush administration, the green technological revolution is coming on faster than ever. But will it come fast enough? And will it suffice in the developing world, where greenhouse pollution is rising fast, and where, frankly, most people spend more time thinking about food and safe water than about global warming?

These are good questions, some of the best, actually, and we're going to take a stab at answering them. For the moment, though, know that the Kyoto Protocol—bitterly hard fought though it's been, and suffering still a deeply uncertain future—is nevertheless only a small preliminary step. Peel away millions of words of criticism and commentary, and this fact remains: Kyoto only imposes emissions caps on the industrialized countries, and these are entirely inadequate to the scale of the problem. Most all the scientists agree: In order to prevent devastating climate shifts worldwide, total global greenhouse gas emissions must soon drop to 60 to 80 percent below their 1990 levels.

What to do? The answer, actually, is clear. We have to build upon Kyoto. We have to construct upon its uncertain foundation a climate protection regime that is both adequate and fair. We have to take the big step beyond Kyoto, and it has to be a step toward a treaty that we can all fight for in good conscience, a treaty that promises an effective global regime even as it opens "environmental space" for the "sustainable development" of the South.

New technologies are essential, but they're not enough. Nor is the problem simply the power of the fossil-fuel cartel, now so clearly displayed in the policies of the U.S. administration. What's really at stake in the greenhouse crisis is the riddle of history: how to find a path to "sustainability" on a planet riven by bitter and explosive national and class divides. Which is where realism comes in, realism and justice. Or, to use the language of the climate negotiations, "equity."

THE SPECTER OF EQUITY

"Equity" isn't a word that brings crowds to their feet. It seems too abstract, too bloodless, too heavily burdened with insincere elite rhetoric. There's no question that when it comes to the language of public activism, "justice" serves far better. It rings with fine moral purpose, echoes in the spirit, and makes for far better slogans. Compare "climate justice" to "climate equity," and you'll see the point. The first wins hands down.

Move, though, from the open air of the street to the considered tones of the conference halls—the halls of the climate negotiations, say—and you move into a different world. The "equity people," suddenly, appear as militants, rushing as they are to make the links, to draw the lines, to demand an honest reckoning. They're the ones insisting that, desperate though the battle for the Kyoto Protocol may still be, we nevertheless have to talk about the future, and soon.

Meanwhile, the "equity agenda," evolving in the climate negotiations and in all the chatter and research that underlies them, is reaching conclusions of terrific importance, conclusions that, strangely, are all but unknown by the larger global-justice movement. Here's one: The Kyoto Protocol may turn out to be one of the most important economic treaties of all time. Here's another: Despite all the attention the global-justice movement has lavished on international trade, it's only within the climate battle that the South really has the power to demand fair terms, and it's unwise to assume that Southern negotiators are unaware of this.

This is a new day, and it demands a rethinking. To that end, *Dead Heat* is designed as an *argument* for a just climate policy, one intended to dispel stale airs, unveil dangerous truths, invite strategic discussion, and, ultimately, expand the global-justice agenda and open doors to new kinds of International alliances. So the first thing to know about this rather grand aspiration is that it may actually be achievable. The threat of rapid climate change is a terrifying one, and there are many among the elites who know this quite as well as we do. Given this, and given 9/11, the global economic gridlock, the suddenly manifest possibility of large-scale war, and all the other rumblings of historical realignment now visible around us, it's clear that North-South relations are changing fast, and that in these changes there are vast opportunities. The question is what we will make of them.

Begin, then, with the specter of equity, now haunting the halls of the UN's climate negotiations. For here, at least, equity is indeed another name for justice. And justice is the thickening form at the center of all those torpid, fluorescent-lit spaces where the strange theater of "environmental governance" has its endless engagements. Only fools imagine that they can ignore it much longer.

REALISM, AS WE KNOW IT

Realism is the favorite pose of the powerful, and some of them even take it seriously. When they do, perhaps late at night, alone with a bottle of old Scotch and a copy of the *Financial Times*, they must eventually find this cold conclusion staring them hard in the eye: If stabilizing the climate at a "safe" level is your goal, global economic apartheid will not achieve it. To make peace within a limited world, you must offer a just peace. There must be justice for nature, and there must, above all, be an attractive future for the poor.

Following the Global Scenarios Group,[1] we'll name the alternative to justice as a "Fortress World" in which the wealthy few do whatever they must do, and become whatever they must become, to maintain their privilege. Skip to chapter 9 for details, or just note for now that the Fortress World, the New Imperium, Global Apartheid—call it what you will—would be as unstable as it would be grim. It would mean that the fossil-fuel cartel and its courtiers remain in power for decades longer, even as the solar/hydrogen economy strains to be born; it would mean endless war and all its pathologies, and the thickening of the cage already visible

Tom Athanasiou and Paul Baer

around us; it would mean a desperate kind of "development" in which vast new plumes of pollution rise from the cities of the South, even as the casualty lists grow ever longer and the greenhouse gas concentration in the atmosphere rises beyond any plausibly "safe" level.

And it would not last.

True realism, not the "crackpot realism"[2] of the Fortress World but the geoecological realism now being born in the ranks and journals and think tanks of the global-justice movement, points in a different direction. It says that "sustainability" must be global, that it must be democratic, that it must work for the poor and the aspiring as well as for the rich and the comfortable, that the alternative is unacceptable, and that we don't have much time. China, Indonesia, Brazil, India—the "big poor countries"—make this point quite as clearly as the newly visible Muslim world. They will be players, and their people will not willingly accept a future that relegates them forever to paths of poverty and bitterness.

A word on Washington: Ever since the U.S. Senate's infamous 1997 Byrd-Hagel resolution, which denounced the emerging Kyoto Protocol and called instead for a climate treaty that included caps on developing-world emissions, "obstructionist" politicians in the United States (echoed in Australia, Canada, and elsewhere) have disguised their hostility to any meaningful climate treaty as a plea for a "global solution to a global problem." Kyoto, they argued, was the wrong approach, and it would penalize its own signatories. To justify their rejection of its targets, they've (correctly) pointed out that the Kyoto Protocol would make only a tiny dent in the projected temperature change, and then gone on to insist that energy-intensive industries would simply move to developing countries, costing jobs in the industrialized countries without reducing greenhouse emissions.

Are we, then, agreeing with them?

Not at all. Despite all the compromises that have been necessary to save it, and inadequate as it is, we support Kyoto. Indeed, Kyoto's inadequacy may well have been a precondition of its survival; had it been stronger, its friends in Europe and the South may not have been able to save it from the administration of Bush-the-Younger, the ExxonMobil lobbyists, the Saudis, and all the rest of the carbon cartel. And Kyoto, as its supporters have long argued, really is only a first step, while this book is about the second, and the third, and the path that, one way or another, these steps must carve through the future.

Forget, for a moment, the mess in Washington. Think globally. And note that Beijing and Delhi also have their politicians, and that they've long insisted that the first move belongs to the North. And who can honestly disagree? The industrial revolution was powered by coal, and later by oil, and thus the North (and, yes, this includes Australia) emit-

Tom Athanasiou and Paul Baer

ted the bulk of the gases that today feed chains of storms, floods, heat waves, and droughts around the world. This is, as they say, history, as is the process by which the North became wealthy and the rest of the world—in contrast and often absolutely—became poor. And it's the basis for our belief that the South will never accept a climate deal in which this monumental grab of atmospheric space is made permanent, no future in which the South restricts its emissions only to see the North, whether by guile, inertia, or sheer blind power, continue emitting massive plumes of greenhouse pollution, as if it were alone in an open world.

Try to look at figure 1 below—showing per capita emissions of carbon dioxide (CO_2) from fossil energy and industry—through Southern eyes. Doing so should make it more than clear why the South expects its own emissions to rise, even while Northern emissions fall.

This figure, like most of this book, and most of the negotiations, focuses on carbon dioxide emissions from energy and industry. For in spite of the role of other greenhouse pollutants (see chapter 2), carbon in the form of carbon dioxide is by far the most important. Though it is now responsible for only about half of the warming, CO_2 has a long atmospheric half-life, and will thus become a larger and larger fraction of greenhouse pollution. And since CO_2 emissions from deforestation are expected to level off and decline (hopefully before the forests are all gone), it's carbon pollution from coal, oil and

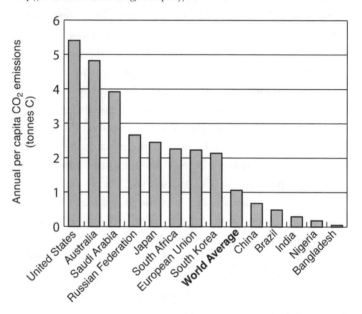

FIGURE 1: 1998 per capita CO_2 emissions from fossil fuel use and cement production for selected countries and regions. Data originally from the Carbon Dioxide Information and Analysis Center (CDIAC), downloaded from the World Bank at http://devdata.worldbank.org/data-query/

gas—the fuels of traditional industrialization—that the global climate protection regime must effectively reduce.

THE PAST AND THE FUTURE

After several years of gestation, the climate treaty was born at the Earth Summit in 1992, when the North, hot under a

Tom Athanasiou and Paul Baer

world-class media glare, promised the money and technology needed to launch a new age of equitable and sustainable development. This was Agenda 21, and the "Rio bargain," and while it fell far short of global justice, or even equality of opportunity, it was really all that was on offer. And it came to very little indeed. With neoliberalism and post–Cold War triumphalism setting the stage, post-Rio aid levels dropped fast, and even the North's promise of technical assistance in areas such as inventorying carbon emissions came to naught. By 1995, when the climate talks began in earnest in Berlin, the "spirit of Rio" was badly frayed, and tension high. (The Bangladeshi Atiq Rahman captured the spirit of Berlin thus: "If climatic change makes our country uninhabitable, we will march with our wet feet into your living rooms.") Decorum was maintained, but only by the adoption of the Berlin Mandate, in which *the industrialized countries pledged to make the first move by accepting, at least for a while, binding greenhouse gas emissions reductions from which the South would be exempt.* This mandate—declaring that the rich would go first—and not the diluted protocol that emerged from the sieve of the climate conventions in Kyoto, The Hague, Bonn, and finally Marrakech, was the key that opened the future of the climate negotiations.

And despite all, despite even the Bush administration, Kyoto may be enough to fulfill the Berlin Mandate. And it had better be. Because flawed though Kyoto is, *there's no good reason to believe that a better treaty was possible in the past, or, indeed, that a better one will be possible in the future if Kyoto is derailed.* If, however, Kyoto is ratified by enough countries,[3] the debate over the terms of the second "commitment period"[4] will begin in earnest, and with it the debates over equity, adequacy, and the next step. *And this time it is extremely unlikely that anything short of a rights-based global treaty will suffice.*

Meanwhile, the politics will be extreme. For if Kyoto becomes law, we'll see an immediate push—by European politicians as well as the United States—for the expansion of the Kyoto framework (and its market-based "flexibility mechanisms") to include ad hoc emissions limitation "commitments" by at least the largest developing countries. And many of our environmentalist friends will argue that these are absolutely necessary if we're to bring the United States into the climate regime (after Bush is out of office, of course). They'll say that this is the only way to meet the terms that Senators Byrd and Hagel have declared that "fairness" demands, and thus to break the anti-Kyoto coalition. And they'll insist, perhaps against their own true wishes, that climate equity, or even sketching a path to climate equity, will just have to wait.

We'll return to all this later. But note, for now, that countries can't really enter the mainstream of the Kyoto system unless they accept emissions limitation targets. *After all, it's only by coming in under your targets that you can have something to trade.* This is a critical problem, for Southern negotiators have always bridled at such targets, and it's easy to see why. After all, how can developing countries accept limits on their greenhouse gas emissions if their development demands that their energy use—and thus their emissions—continue to increase?

It's a pretty problem. And a global accord based on per capita emissions allocations (our goal) is not the only solution being put forth, not by a long shot. The alternative (to massively oversimplify) is an incremental approach in which some developing countries would accept not *reduction targets* (like those that Kyoto imposes on its signatories) but, rather, *growth targets* that allow their emissions to increase within bounds. A variety of arcane, more or less ad hoc methods for calculating such targets are currently being auditioned in the think tanks and ministries of the North.[5] And if there's anything of which we can be sure, it's that the South will soon come under pressure to accept one or another of them.

Which would, at first glance, be fine. After all, if developing-country emissions continue to increase at anything like their current rates, a catastrophic degree of climate change is a foregone conclusion. And it would be the developing countries themselves, their peoples and their lands, that would likely suffer the greatest devastation.

There's one problem, though: Unless we're all very careful, we could wind up grandfathering the sky over to those who were the biggest polluters in 1990, Kyoto's year zero. We could, in other words, leave the North with a vastly disproportionate share of the global greenhouse gas budget, even as that budget shrinks, as the science says it must. In this case, to put it baldly, we'd leave the developing world without any atmospheric space to develop into.

The South is not at all unrealistic to fear that grandfathering would be the outcome of an ad hoc, incremental approach. Moreover, such an approach would pose an imme-

The Environmental Space Bench, courtesy of the Centre for Science and Environment, New Delhi, India. RUSTAM VANIA/CSE

Tom Athanasiou and Paul Baer

diate danger: If the key Southern countries think it risks acquiescence to grandfathering, then they, along with their allies in Europe and among the greens, will generally if not unanimously resist it. The result would be deadlock or, in the best case, a begrudging agreement that's altogether inadequate to the challenge at hand.

This is actually a likely scenario, as the debate over developing-country commitments heats up later in the decade. So cast your eye a few years ahead. Kyoto will have entered into force, and its institutions will be taking shape. Evolution will seem the logical way forward, but the South will be resisting. And many people, we among them, will be arguing that the time has come for a real solution, one that launches a global regime in which national emissions budgets are set in an adequate, fair, and comprehensible manner. Moreover, climate, by that time, will be widely recognized as both a global crisis and a crucial front in the battle for global justice.

The time is coming, so note the sad state of the current debate, and how poorly it prepares us. The nub of the equity argument is that to be effective, the climate-protection regime must both be fair and be seen to be fair, and that nothing but a per capita framework will do. Many climate policy experts, when pressed, will actually agree, but only "in the long run." They'll invariably argue that "equity" cannot take center stage, not yet; that per capita rights are a "nonstarter." And they'll conclude with their decisive point, that the climate accord can't reasonably be asked to remedy all the inequities in the world.

The point must be conceded. But here is its converse, equally true: A workable climate treaty must at least help to break the deadlock between the affluent and the aspiring. It must satisfy the claims of common-sense justice and accommodate the South's all-too-legitimate aspirations for "development." It cannot treat environmental justice as a minor rhetorical complication, and far from the prevailing spirit of "global environmental management," in which Northern greens have often teamed with institutions such as the World Bank and the International Monetary Fund to reform the practices of the poor, it must look first of all to reform in the North, and to the pathologies of affluence. It must, finally, be a constitutional, "rights of man" kind of treaty, one in which we affirm that we all, however proud or humble we may be, have the same ultimate claim to the atmospheric commons.

Tom Athanasiou and Paul Baer

CHAPTER 2

THE SCIENCE CHAPTER

One of the greatest obstacles to stopping climate change is the difficulty of understanding it, and one of the greatest obstacles to understanding it is the confusing tangle in which climate politics and climate science are intertwined. Nevertheless, we've little choice but to try to sort out the threads.

Severe climate change—and with it chains of storms, floods, heat waves, droughts, and even cold snaps—is now virtually inevitable, as is widespread ecological destruction, extinction, and human suffering. Continued dithering will predictably lead to climatic instability on a truly terrifying scale. We actually *know* this, even though it hasn't happened yet, and it takes a fairly good grasp of the science to understand why. Further, there are significant ways in which we're *entirely* dependent on science to understand either the climate problem or the demands it makes upon our responses. Frankly, some strategies will work and others won't, and we need a good grasp of both the science and the politics to tell the difference.

Our goal here is not a comprehensive one. We suffer no illusion that we can summarize climate science as a whole.

But we do think that we've distilled out the part of the science that bears most immediately on the *core problem of drawing the line*. Should the allowed maximum of CO_2 pollution in the atmosphere be 550 parts per million (ppm), 450 ppm, or some other (hopefully lower) figure? Or should we take an entirely different approach and try to cap global temperature change itself, rather than CO_2 pollution? And what must we know about the kinds of impacts and instabilities that can be expected at any given level? As we will show, the current science tells us that we should care deeply about these questions, and about the very specific calculations they inevitably involve, for behind these calculations lie choices of life and death for millions of humans, and survival or extinction for thousands of species. So bear with us, if you will, as we discuss concentration caps, radiative forcing, climate sensitivity, and increased climatic variability. We'll try to make this as painless as we can.

THE IPCC'S ASSESSMENTS

The global-warming crisis has given rise to a unique scientific body, the Intergovernmental Panel on Climate Change (IPCC), which brings together thousands of scientists from around the world in a tightly focused process designed to provide continually updated assessments of the threat, the science that allows us to know the threat, and the uncertainty of that science.[1]

Tom Athanasiou and Paul Baer

The "skeptics," of course, have attacked the integrity of the IPCC, but a recent report by the prestigious U.S. National Academy of Sciences—one the Bush administration itself requested!—strongly endorsed both the IPCC's process and its assessments, and denied that they have become politicized.[2] Indeed, *Climate Change Science: An Analysis of Some Key Questions*, as the NAS report is called, forced the administration into a position in which it either had to admit the seriousness of the climate change problem or be widely and visibly seen as turning its back on science. Which is no doubt why, only days after its release, on June 11, 2001, George W. Bush gave the infamous speech in which he conceded that "the National Academy of Sciences indicates that the [temperature] increase is due in large part to human activity," but nevertheless went on to repudiate the Kyoto Protocol as "fatally flawed."

It's important to understand that this move—from "the science is uncertain" to "the treaty is flawed"—is being forced upon the fossil-fuel cartel as the science advances to the point where "climate skepticism" has come to remind even mainstream observers of a tobacco company public relations campaign.

The Bush people hate the IPCC, and there are reasons why. The IPCC's tasks include the preparation of "state of the science" assessments every five years, and these reports have played a crucial role in making the situation clear. In what follows, we draw heavily on the IPCC's Third Assessment Report, completed in 2000 and published in 2001.[3]

GREENHOUSE GAS BASICS

Human-generated (anthropogenic) greenhouse gases such as CO_2 warm the Earth by trapping additional solar radiation within the atmosphere. Many gases besides CO_2 act as greenhouse gases, most notably water vapor, but also methane (CH_4), nitrous oxide (N_2O), and fluorocarbons (some of which, the chlorofluorocarbons, or CFCs, are being phased out due to their ozone-destroying properties). So understand, first of all, that the greenhouse effect is natural; before humans started increasing the concentrations of these gases in the atmosphere, these gases worked together to create a "blanket" that keeps the Earth far warmer and more habitable than it would otherwise have been.

Unfortunately, once the industrial revolution began to leverage the cheap and easy energy of fossil fuels and launched the age of industrialization, the atmospheric concentrations of the greenhouse gases—now greenhouse pollutants—began to rise rapidly. Since the 1700s, CO_2 alone has increased from about 275 ppm to over 370 ppm, and it continues to rise, at about 1.5 ppm per year. The significance of this increase is measured first in terms of "radiative forcing"—which in this context is a measure of the amount of

Tom Athanasiou and Paul Baer

additional energy trapped or reflected by the greenhouse gases that humans have added to the atmosphere.

Positive radiative forcing means an increase in the solar energy absorbed from the sun, and it produces the kinds of changes that you'd expect—generally warmer temperatures, and changes in the patterns and variability of the weather. The increase in radiative forcing attributable to humans (for just CO_2) is so far about 1.4 watts per square meter, a substantial fraction of the 3.7 watts that would be expected from a doubling of CO_2 concentrations (to 550 ppm from the preindustrial level of 275 ppm). And since the non-CO_2 greenhouse gases add about 1 watt per square meter, you can see that they are also extremely significant.[4]

Note, though, that the warming effect of these gases is partially offset by a negative radiative forcing (cooling effect) caused by other pollutants called aerosols,[5] particularly sulfur compounds produced by combustion. This is an important point because many of these cooling pollutants are extremely dangerous to human health, particularly in the local communities where they concentrate in what are called hot spots. They also lead to acid rain and they must be quickly eliminated. Doing so, however, will mean a larger net positive radiative forcing, and, in fact, this positive forcing is inevitable—unless we continue to indefinitely dump aerosols into the air—because they have much shorter atmospheric lifetimes than the major greenhouse gases.

How much should we care about one or two watts per square meter? It depends on another key link in the chain: the overall responsiveness of the climate to radiative forcing. Climate scientists call this the "climate sensitivity," and by this term they mean *the estimated long-run increase in global average surface temperature that would be caused by a doubling of* CO_2 *from preindustrial levels*—that is, its rise to 550 ppm, a level we could easily reach in the next fifty to a hundred years. In both the Second (1995) and Third (2000) Assessment Reports, the IPCC estimated this factor to be in a range between 1.5°C and 4.5°C, with a best estimate of 2.5°C cited by the Second Assessment Report. (To compare, the Earth is now only about 5°C warmer than it was 18,000 years ago at the peak of the last ice age.)

THE IMPACTS: PRESENT AND FUTURE

Let's step back, glance at the evidence that humans have already changed the climate, and consider the impacts that we can expect to see as the changes intensify.

Start with the evidence that the Earth is actually warming. This could be a very long list, but if we just pick a few of the most significant bits (most of them from the IPCC's Third Assessment Report), we have:

➤ Since the industrial revolution, the global average surface temperature has increased by about 0.6°C. And that's the *average*: The temperature is increasing much more quickly near the poles, and many scientists now expect the Arctic ice cover to be almost entirely gone by 2080.[6]

➤ Globally, the 1990s were the warmest decade and 1998 the warmest year since 1861. So far, 2001 was the second-hottest year overall, though its winter took first place. And the records just keep on coming!

➤ Ongoing changes in sea level, snow cover, ice extent, and precipitation are consistent with a warming climate near the Earth's surface. For example, there has been a wide-spread retreat of non-polar mountain glaciers during the twentieth century.

➤ The rising costs of weather damage, much of it caused by recent increases in floods and droughts, is itself a good indicator of increasing "climatic variability." (The more farsighted insurance companies are becoming extremely worried.)

➤ The warming in the twentieth century is the largest of any century during the past thousand years, which can be easily seen in the famous and somewhat terrifying thousand-year temperature chart shown in figure 2 (it's known, by the way, as "the hockey stick").

Here you can see the Earth's average Northern Hemispheric temperature anomalies—the annual difference from the

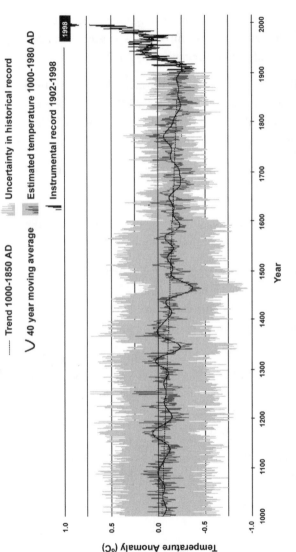

FIGURE 2: Global mean surface temperature record for the last thousand years, measured as annual anomaly (variation from the 1902 to 1980 mean annual temperature). Adapted from Mann, M. E., R. S. Bradley, and M. K. Hughes. 1999. Northern Hemisphere Temperatures During the Past Millenium: Inferences, Uncertainties, and Limitations. *Geophysical Research Letters* 26:759–62.

twentieth-century mean—over the last thousand years (the dark, meandering line), a decreasing margin of error (the initially broad but thinning shaded area that surrounds that line), and a gradual long-term cooling trend that abruptly ended about 1900. There's a lot here, but note the singularity of sudden twentieth-century warming, and note especially the steep vertical slope of the curve in the last decade.

So something is happening. But are humans really causing it? Again, we defer to the IPCC, which in 1996, in their Second Assessment Report, issued the carefully crafted and oft-quoted phrase, "The balance of evidence suggests that there is discernable human influence on global climate." By the time of their Third Assessment Report, published in 2001, the IPCC's prose had solidified, and it told us that "there is new and stronger evidence that most of the warming observed over the last 50 years is attributable to human activities."

The "skeptics," of course, still argue that the observed temperature increase and other signs of climate change haven't been proven to be a result of humanity's greenhouse gas emissions. Is this a reasonable objection? Could the warming so clearly visible in the hockey stick be a result of "natural variability"?

No, it could not. For while the climate varies over a wide variety of time scales—from El Niño cycles of a few years, to glacial cycles of tens of thousands of years, to the even longer cycles of the Earth's deep history—none of these explains the trend so clearly visible in the hockey stick. But basic laboratory physics does. We know that greenhouse gases trap solar energy, and we can measure the rapidly increasing concentrations of these gases with grim precision. Given this, we actually *expect* significant warming, and in much the same pattern that it actually appears (for example, the warming peaks are at the poles and in the winters).

It's not "proof," but it will do.

Uncertainties remain, but they no longer justify doubt that we're changing the climate. In fact, the "skeptics'" traditional comment, in reply to the evidence for human-induced climate change, is that computer models project temperature increases that are *higher* than those actually observed. Alas, with the recent inclusion of cooling aerosols into these models, the projections have converged with the observed record. And now, rabid talk show hosts to the contrary, even George W. Bush cannot safely deny the solidity of the science. So let us, then, just accept it, and go on to the next question: What kind of climate change are we in for?

This one is harder, for the shape of the future will depend on a good deal more than atmospheric science. It will turn as well on the character of our farms, factories, and energy systems; on the nature of our economies and cultures; on the trajectory of "globalization"; on our success in avoiding a descent into hatred and militarism; and, in general, on the *kind of soci-*

Tom Athanasiou and Paul Baer

ety we have. Such things cannot be modeled, not directly. But they can be represented by "proxy" variables such as population, economic growth rates, and quantitative indicators of technological change, and these can be combined with computational models of the atmosphere, land, and ocean to yield suggestive projections of future climate change. The point to remember about such projections, though, is that they're subject to "story-line uncertainty" as well as "scientific uncertainty." You may have a brilliant integrated assessment model,[7] but its results will nevertheless depend not only on assumptions about climate sensitivity and carbon cycling but also on assumptions about, say, the degree of regionalization in the global economy, or the faith that our children will put in technological rather than political realignments.

We can't do much about the climate sensitivity except use the best science to try to estimate it, and in the meantime face the possibility that it'll come in on the high side. The same, however, can't be said about the story line. In fact, the story line—the tale of our common future—is quite literally up for grabs. And it's a damn good thing, because one thing we know for certain is that we don't want to end up where we're currently going: Computer models looking at various plausible scenarios of the future are bringing in warming projections as high as 5.8°C by 2100, and to say that such a warming would be a social and ecological disaster is to strain the limits of understatement.

How bad would it be? Realize, first of all, that average temperature increases are only the beginning of the story. Actual increases will vary greatly around the average, and temperatures are projected to increase more over land, more in the higher latitudes, and much more near the poles. Further, the variability of the climate will also increase, meaning both extreme temperatures (mostly hot but also cold) and large and possibly abrupt changes in the water cycle, and thus severe droughts and floods. The expected rise in sea level (as much, in the worst case, as a full meter in the next hundred years) will have devastating and sometimes apocalyptic impacts on low-lying areas, rendering many small-island states uninhabitable and multiplying the risks that hundreds of millions of coastal residents face from increasingly severe storms and inundations. Most coral reefs will die. Large-scale tropical forest die-offs are likely, and radically increased "food insecurity" (read: starvation) is a near certainty. Rice production in Asia will be hit especially hard. Human and ecological migrations will increase, and with them political and military tension. Surprises are certain.

We could go on here, but if you really want the story of the impacts to come, go to the IPCC itself. Go, in fact, to the Summary for Policymakers of the IPCC's Working Group 2, which focuses on "impacts, adaptation, and vulnerability."[8]

And there's another thing we have to mention: the risk from what the IPCC dryly calls "large-scale discontinuities." There are a number of terrifying possibilities here: a rapid release of carbon and methane now bound within various oceanic and biological "sinks" (which remove carbon from the atmosphere),a sudden large rise in sea level caused by ice melts in Greenland or Antarctica, or the sudden collapse of the "thermohaline circulation"—the large-scale ocean current that moves warm water from the tropics toward the poles, and that warms northern Europe. None of these discontinuities has yet occurred, but the IPCC's reports suggest that they could, perhaps even within the lifetimes of our children. And the fact that they're hard to model should not be taken as a source of solace; indeed, there's always the possibility of true surprises: changes we didn't even know enough to worry about.

Again, the uncertainties associated with increasing greenhouse pollution make it impossible to predict which specific impacts will follow from which concentration levels. Nevertheless, the IPCC has made an effort to produce an initial "vulnerability chart" (figure 3), and it generously repays a bit of study. Note that the horizontal axis measures the change in degrees Centigrade from the preindustrial average.

The chart in figure 3 shows us that:

➤ We're *already* experiencing risks to unique and threatened ecosystems.

➤ The risks of extreme climate events have *already* risen.

FIGURE 3: Potential impacts from climate change with increasing change in global mean surface temperature. Adapted from the IPCC's Third Assessment Report, 2001.

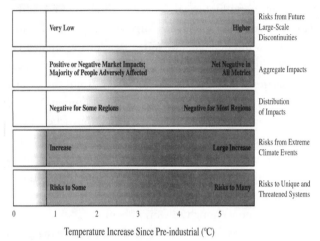

Temperature Increase Since Pre-industrial (°C)

➤ The overall impacts of climate change are *already* negative for some regions and the majority of people.

And it shows that if global average surface temperature rises by more than about 2°C—an extremely real possibility, even a likelihood, given the way things are going—the risks of extreme climate events will show a further large increase, and the impacts on almost all regions and economic sectors will become negative. Further, with a temperature rise of around 3.5°C, or even less if the more pessimistic scientists turn out

Tom Athanasiou and Paul Baer

to be correct, the risk of "large-scale discontinuities" (potential world-historic catastrophes) will become significant.

THE TWO-DEGREE STANDARD

In 1996, the European Environment Council (EEC) decided that the global average surface temperature increase should be held to a maximum of 2°C above the preindustrial level, and that as a consequence the CO_2 concentration had to be held below 550 ppm. Unfortunately, such a 550-ppm concentration limit would only be "safe" if the climate sensitivity turns out to be very low, which (as we will explain) is quite unlikely. Even more unfortunately, the number quickly became a popular one. Soon thereafter, Bill Clinton also announced that his policy was to stabilize at 550 ppm, and, no doubt coincidentally, rumors began to fly that the United States was pressuring the IPCC, then busy drafting its Third Assessment Report, to feature 550 ppm as its principal mitigation scenario.

Fortunately, the pressure failed. And, in fact, the numbers just don't add up. As we will show, a CO_2 level of 550 ppm would almost certainly bring a temperature increase of *far more than 2°C* and lead to horrific levels of destruction. In all likelihood—according to the IPCC's Second Assessment Report—it would be accompanied by significant ecosystem damage and loss of biodiversity ("whole forests may disappear"), significant damage to food production in the most vulnerable parts of the world (60 to 350 million more people at risk of hunger), "significant loss of life" due to indirect health effects, particularly in developing countries, and, of course, a significant increase in sea level.

Look again at the vulnerability chart in figure 3 and understand that just because a group of politicians decides that 550 ppm would be tolerable, this hardly makes it so. Moreover, even if we're very lucky and the climate sensitivity turns out to be so low that 550 ppm would map to a warming of only 2°C, this would still be very grim indeed. Be clear about this. A 2°C warming would be a death sentence for tens of thousands and perhaps millions of people, a commitment to catastrophic losses of species and ecosystems, and, frankly, an invitation to a dangerous new exacerbation of geopolitical and military instability, one that we hardly seem likely to manage with aplomb.

The real point, as Greenpeace has been stressing for years in a series of reports on "the carbon logic," is that we simply must not burn all of the fossil fuels at our disposal, or even all the gas and oil, and that even burning most of them would produce an ecological holocaust. Greenpeace strategists do not imagine that they can avoid the politics of limits, which demand numbers, but unlike the EEC and the Clinton administration, they set their limit not by the realism of the moment but with a close, prudent reading of the science. A warming of 0.6°C has already occurred, and a rise of 1.0°C will be here very soon, but they nevertheless counsel us that "temperature changes above 1.0°C above prein-dustrial levels could bring about rapid and unpredictable changes to ecosystems, leading to large damages."[9]

A global temperature change cap of 1.0°C would be just fine with us; though, in truth, a greater degree of warming may already be locked into the atmospheric system. Indeed, *Dead Heat* is, ultimately, an argument for an equity-based strategy as the best way to make low targets achievable. But what we want to do, right here, is perform a different sort of thought experiment, one in which we bow to pessimism—or is it realism?—and ask just what it would mean to draw the line at 2°C. If that were our goal, but seriously this time, how much greenhouse pollution could we, all of us together, emit in the next few decades? Such a calculation is necessarily uncertain, but if we did it honestly, what would it show?

To answer this question, we're going to look again at the link between greenhouse gas concentrations and tempera-ture increase. In fact, we're going to show you the most dif-ficult chart in this book. Take a look at figure 4: first, at the three diagonal lines, each of which represents (and simpli-fies—in reality these are not straight lines) one of the IPCC's classic low (1.5°C), best (2.5°C), and high (4.5°C) estimates for the Earth's overall climate sensitivity. For each of these values, one of the diagonals shows the relationship between the atmosphere's CO_2 concentration and the expected tem-perature increase.

FIGURE 4: Relationship between CO_2 concentration and equilibrium increase in global mean surface temperature for low, best, and high estimates of climate sensitivity (ΔT).

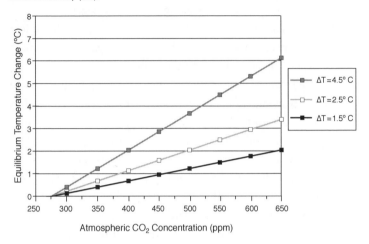

THE LOW-SENSITIVITY CASE

We're not going to say much about this one, because the case for low climate sensitivity is becoming increasingly implausible. Suffice it to say that 1.5°C, the low end of the IPCC's standard range of climate-sensitivity values, dates back to 1990's First Assessment Report, and that few current models corroborate it.

THE IPCC'S "BEST" ESTIMATE, WITH NO OTHER GASES

The 2.5°C climate-sensitivity diagonal, as you can see in

Tom Athanasiou and Paul Baer

figure 4, crosses the 2°C temperature-change line when the projected CO_2 concentration reaches 500 ppm (as against the preindustrial level of 275 ppm and the current level of above 370 ppm). What this tells us is that if we were willing to accept the social and ecological consequences of a 2°C increase (we are not), and if the climate sensitivity is 2.5°C (which could easily be low), and if there were no "other" greenhouse gases (there are), we wouldn't want to exceed a CO_2 concentration of 500 ppm.

THE HIGH-SENSITIVITY CASE

The IPCC puts the high end of the likely climate-sensitivity range at 4.5°C, but the current scientific trend is to considerably raise that upper limit. Still, let's go with the IPCC's *conservative* upper estimate of 4.5°C. If this turns out to be the climate sensitivity, then follow the 4.5°C climate-sensitive diagonal to where it crosses the 2°C temperature-change line, at 400 ppm. This is, as it happens, only 30 ppm above today's CO_2 concentration level.

THE ADDITION OF NON-CO2 GASES

Now step into the real world, and consider the *additional* radiative forcing from non-CO_2 gases. The IPCC's scenarios put the net non-CO_2 forcing in 2050 (including the cooling from sulfate and other aerosols) at between 0.3 and 1.2 watts per square meter, equivalent to about 20 to 75 ppm of CO_2. Taking a midrange value of 50 ppm and subtracting it from the numbers above, we can get a good estimate of the CO_2 concentrations that, for each of our climate-sensitivity estimates, is actually likely to correspond to a warming of 2°C.

The results are quite frightening. For example, if the climate sensitivity turns out to be 2.5°C (which is probably low), we'd get a 2°C increase at a CO_2 concentration of about 450 ppm (rather than at 500 ppm if we ignore non-CO_2 gases). And if it comes in at 4.5°C (which is probably high), the 2°C point would come at 350 ppm, rather than at 400 ppm; this would mean we're already over the 2°C line, though for a variety of reasons (the absorption of heat by the oceans, and the fact that past emissions have "locked in" but not yet delivered an unknown amount of future warming) we don't know it yet.

THE PUNCH LINE

This is all very rough, and *very* simplified, but believe it or not, we haven't done violence to the facts. We're using the IPCC's numbers, and if we've had to connect the dots ourselves, it's only because the scientists, for their own reasons—bad and good—are reticent to do so.[10]

When it comes to the "real" climate sensitivity, note that the IPCC's Third Assessment Report, while still reporting the 1.5 to 4.5°C range, doesn't use its old figure of 2.5°C as

Tom Athanasiou and Paul Baer

the "best estimate" of climate sensitivity. And note, too, that recent studies by climate modelers and recent estimates drawn from the ice-core record both suggest that the median estimate is likely to be closer to 3.5°C, with significant possibilities of 5°C or higher. [11]

And if the climate sensitivity turns out to be 3.5°C, then the CO_2 concentration target corresponding to a warming of 2°C will, when corrected for the non-CO_2 gases, be about 400 ppm, which is coming up in twenty years or less.

The bottom line is not a pretty one. Both the climate system and the civilization we've built within it have a great deal of inertia and will be difficult to turn. The atmospheric CO_2 concentration is increasing year by year, and this increase isn't going to be easy to stop. And remember, please, that none of the numbers here include the possibility of self-reinforcing terrestrial or oceanic feedbacks—the dreaded "non-linearities" that, set off by the heat-related dieback of tropical forests or the release of methane currently bound up in oceanic "sinks," would take all these fine projections off the table and leave us, instead, in climate hell.

In short, we haven't got much time.

CHAPTER 3

FROM TEMPERATURE TARGETS
TO EMISSIONS BUDGETS

The last chapter argued that a temperature target should drive global climate strategy, and that this target should itself be defined by "acceptable" impacts. The obvious question is, acceptable to whom? But even leaving it aside, uncertainty and the need for a precautionary approach complicated the analysis at every point. Still, we found a bottom line in the fact that a temperature target allowed us to set a concentration target, and we're now going to take the next step and introduce the notion of a global emissions budget. And, in fact, this budget—how much CO_2 and other greenhouse gases (GHGs) we can "safely" emit—is the real issue; ultimately, it's only units of greenhouse gases, not units of temperature increase or carbon concentration or sea-level rise, that can be regulated.

Another word, though, about our temperature target.

We'd like nothing better than to advocate a temperature-change target of 1°C, and (following Greenpeace's "carbon logic,") insist that no greater warming can be tolerated. But we're talking instead about 2°C. Why? Because we think that, perhaps strangely, doing so allows us to better underscore the severity of the situation: If we can take the real uncertainty into account (as opposed to the uncertainty imagined by the "skeptics"), cut ourselves 2°C of slack, and nevertheless conclude that we're in trouble, we have a pretty strong case.

In that same spirit, let's assume that we're going to be lucky, and that the Earth's climate sensitivity will turn out to be only 2.5°C (lower than many scientists expect). What this would mean is that the atmospheric concentration of carbon dioxide can rise all the way to 450 ppm while still allowing us to reach a "soft-landing corridor" in which emissions don't force the warming above 2°C.

How, then, are we to make it to a 450-ppm corridor?

AN (OPTIMISTIC) SOFT-LANDING CORRIDOR

Figure 5, adapted from the IPCC's Third Assessment Report, shows year-by-year projections of allowable emissions under a 450-ppm CO_2 stabilization corridor, as calculated by one highly respected carbon cycle model.[1] There are three emissions paths here, with the area between the highest and lowest shaded to show the "450 corridor." The width of the corridor reflects the uncertainty of carbon-cycle science. (This, by the way, is an uncertainty we haven't discussed; the questions here have to do with how and how quickly natural processes remove carbon from the atmospheric system.[2])

FIGURE 5: A 450-ppm stabilization corridor, the width of which reflects uncertainty about the global carbon cycle. Adapted from the IPCC's Third Assessment Report, 2001.

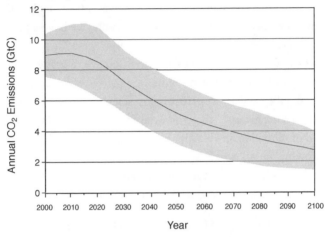

Uncertainty is the whole point of the corridor metaphor. The two edges of the corridor contain the uncertainty and allow us to clearly visualize it. This is important because scientific uncertainty is incessantly invoked by the carbon cartel and its pitchmen as an excuse for inaction, and thus we have to understand it as clearly as possible. Carbon-cycle uncertainty, by the way, isn't going to be very useful to ExxonMobil. There's just no way to use it to make our predicament look any better.

In looking at figure 5, focus on the areas under the curves. It's these that show, for each of the three cases, what the

Tom Athanasiou and Paul Baer

allowable cumulative global emissions would be over the next hundred years. *Note that even if we follow the highest and most permissive path, global carbon emissions must peak in less than twenty years, and then head steeply down.* And if we want to err on the precautionary side, we shouldn't just look at the top curve. If the low-emissions path turns out to better describe the carbon cycle's behavior, then there's a whole lot less atmospheric space to go around. Indeed, we may already be overshooting the 450-ppm path, since global emissions are continuing to rise, rather than falling, as the lower path dictates.

Even in the middle path, which just about everyone seems to use as their soft-landing "marker scenario," global emissions must be far, far lower in fifty years than they are today, and this must be true even as the developing world continues to, well, develop. And this is all to meet a 450-ppm target, which we may someday recognize as substantially too high.

Again, allowable cumulative global emissions in the next hundred years are shown here as the areas under the curves, and these turn out to be roughly 750 gigatonnes of carbon (GtC) on the high path, 550 GtC in the middle, and 375 GtC on the low path. With current emissions at about 8 GtC per year, the low path would see us use the entire available carbon budget in as little as forty-five years, and that's if annual global emissions don't grow at all.

Despite all uncertainty, this cumulative limit tells us something we very much need to know: how much "environmental space" is actually available. The metaphor is telling: In the case of the atmosphere, we are almost literally "filling up the space," leaving less to our children and grandchildren. Further, the North has already used up far more than its fair share.[3]

Look back at the middle curve. This is a plot of the numbers that the IPCC dryly reports in its Third Assessment Report, when it tells us that stabilization at 450 ppm requires that global emissions be reduced to below current levels by 2050, and to 3 GtC annually by the end of the century. You'll be seeing this curve again, and not just in this book, so look hard at the trajectory it describes. Remember that this (medium) 450-ppm path requires that global emissions peak and then start dropping in less than fifteen years, and that this is quite impossible if emissions from the South continue to increase along the "business as usual" path.

"BUSINESS AS USUAL" AND THE IPCC'S SRES SCENARIOS

We know that things must change, but how much? It's hard to say, and for a very specific reason—we can't discuss the size and scope of the needed changes unless we have a sense of the "baseline," the "do nothing" case against which they can be compared. And here the issue is not just uncertainty but also ideology. Economic models of the costs of meeting

Tom Athanasiou and Paul Baer

emissions targets are *extremely* sensitive to their baseline scenario assumptions, and their results cannot honestly be considered apart from them. Indeed, if you want to train yourself to see through the political agendas hidden in economic and social scenarios, look first for their baseline assumptions.

Deconstruction, of course, only gets us so far. The question is what we should do. In order to model any given social or policy change, we have to compare it to a baseline, yet it's difficult to predict economic or technological changes even one or two decades into the future. How can we talk, then, about a "business-as-usual" future that lasts for half or even a full century?

In fact, there's no way to decisively extrapolate present trends. The actual "present" is a dynamic one, with a large variety of our world's key features changing simultaneously, and the interplay of all these changes is quite literally impossible to predict. The growth of simple, measurable quantities like energy use and carbon emissions is the product of complex changes in economic structure, energy mix, and other unpredictable and contingent factors. Historical events such as the political transition in Eastern Europe, the conversion of the British energy system from coal to gas as a result of Thatcher's desire to break the unions, and the nuclear accidents at Three Mile Island and Chernobyl have all changed global energy and emissions patterns. The history of the prediction of energy use should be humbling in this regard.[4]

Which is to say that any "baseline" scenario—in which the present is extrapolated into the future to find out what will happen "if we do nothing"—must necessarily reflect beliefs and politics and values, as well as facts. Which is, in turn, to say that "business-as-usual" scenarios generally reflect the desires, or at least the beliefs, of those who want to do business as usual.

It's a real problem. Fortunately, the IPCC has recently engaged it by producing new kinds of scenarios and working hard to replace the business-as-usual approach with one based on a set of "story lines," a set, that is, of possible futures with distinct and explicit political and technological assumptions. These are the SRES scenarios,[5] and they're a long story that we're not going to tell right here., not in any detail.

We are, though, going to offer you this one picture, and a bit of commentary. Figure 6 shows the four SRES "scenario families" arranged into four quadrants, with globalization and regionalization defining one axis and political/cultural emphasis—from "material wealth" to "sustainability and equity"—defining the other. It also shows that the high-growth, high-technology quadrant (A1) has three "subfamilies," one with a continued emphasis on fossil fuels, another with a sharp turn toward renewables, and the third balanced between the two. And, just to make the overall situa-

Tom Athanasiou and Paul Baer

FIGURE 6: Sample scenarios from IPCC's four SRES families, shown against a 450-ppm trajectory (dotted line). See text for discussion of axes. Adapted from Berk, M., J. van Minnen, B. Metz, and W. Moomaw. 2001. *Keeping our Options Open: Results from the Cool Project.* Dutch National Institute for Public Health and the Environment (RIVM), Bilthoven.

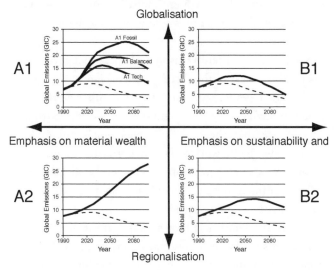

tion crystal clear, it overlays a "best fit" 450-ppm soft-landing path (like the medium-case path discussed above) on each quadrant.[6]

Even without knowing the details behind these charts, you might be able to see how nicely they outline our predicament. Look at the highest curve in the A1 quadrant, the one in which fossil-fuel neoliberalism continues as the dominant theme, and note that, bad as it is, it's not actually the worst case. That honor goes to the A2 world, in which "material wealth" remains the foundation of human culture, but the drive toward globalization sputters into a regional standoff in which globalization fails—in an extremely grim manner in which overall inequality continues to increase.

Note, too, that there's reason for hope, if not actual optimism. The lowest of the three A1 curves (green technology revolution), together with the scenario families on the "sustainability and equity" side, and the "B1 world" in particular, indicate that there are still open roads out of here. Importantly, the B1 world is one that prioritizes equity, sustainability, *and* globalization, though not globalization as we know it today. It's a vision of North-South cooperation of a new and desperately needed kind.

All the SRES curves fall too slowly to meet the 450 corridor, but this isn't as bad as it may seem, for the SRES world is one of "nonintervention." It assumes no climate policy at all, seeking instead to substitute a number of plausible "story lines" for one naive "baseline" case, and to see what would happen in each instance. In this sense *all* the SRES futures are business-as-usual futures, and this, of course, makes them entirely implausible, for there will be a climate policy. The funny thing, though, is that SRES is all the more instructive for being climate-policy free. What its story-line approach finally shows is that the overall shape of the

Tom Athanasiou and Paul Baer

future—the logic of our economies and the development of our "values"—is likely to have a greater impact on the future climate than any climate policy itself. Unless, and this is an important proviso, that policy were to change, or help to change, the overall course of history.

DECONSTRUCTING "BUSINESS AS USUAL"

Although business-as-usual (BAU) scenarios can't be taken as projections of a "most likely" future, there are reasons to explore them more carefully. For one thing, elite spokespeople make great efforts to present variants of the existing world as the most likely and "natural" futures. And these clearly represent the desires of what we will call "the party of business as usual," the future that the broad center of the political class would like to see, if (for example) climate change weren't a problem.

To be explicit, the typical BAU future includes high economic growth, but higher in the South than in the North, so that the North can continue to get wealthier (in absolute terms) while emerging markets boom and global inequality decreases. And, following current trends, BAU includes substantial increases in energy efficiency. And, to be clear, BAU would lead to a climatic disaster.

There's a twist here. In spite of the IPCC's insistence that there is no "central case" scenario—the SRES report explicitly states that no scenario family is "any more likely than any other"[7]—it's not hard to pick out the de facto BAU path. It's the A1 "balanced" (A1B) case above (the middle curve in the upper left quadrant), in which development in the South follows the gradually more efficient path of the North. More specifically, A1B sports rapid economic growth, a "balanced" energy mix (which includes substantial increases in zero- or low-carbon energy as well as continued dependence on fossil energy), a 2050 emissions peak, and, subsequently, a population decline. It's worth a closer look.

Figure 7 shows one representation of the A1B path, the one specified by the IPCC as a "marker" for use by other modeling groups. It shows the projected growth path of CO_2 emissions (including emissions from forest cover and other land use changes) divided into North and South, with global emissions (the sum of North and South) tracking the top of the lighter shaded area. And, just to scare you (though not as much as we could), we've overlaid (as per the discussion above) a medium-case 450-ppm soft-landing path.

What these curves show is that global emissions crack through the "safe" 450-ppm path by about 2005, and that *Southern emissions alone* (the top edge of the darker shaded area) exceed that path by about 2020. What it doesn't show is that the more global emissions increase above the 450 path, the faster and farther they'll have to fall to return to it, and that if they follow the A1B trajectory, we can a expect a catastrophic temperature change upwards of 4˚C.

Tom Athanasiou and Paul Baer

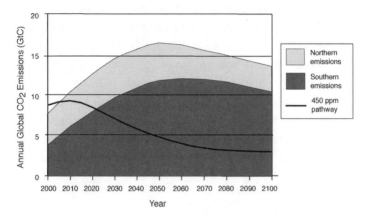

FIGURE 7: Annual CO_2 emissions North and South under IPCC's A1 "balanced" scenario, plotted against a midrange 450-ppm stabilization pathway. Data from the IPCC's *Special Report on Emissions Scenarios*, 1999.

And this, dear friends, finally explains why the situation is so dire, and why half-measures will not avail us. The 450-ppm path—dangerous though it is—defines an emissions budget for the coming century that, clearly, is far too small to accommodate "business as usual." To make it down to the 450 path, we'll have to win major changes in the trajectories of not only the developed North but the developing South as well.

Or put it another way: The North could meet its Kyoto targets, and, all else being equal, climate change would still reach highly dangerous levels unless the curve of Southern emissions also bends down, sharply, and soon. But why—

think "real world" here—would the South strain to make such a turn unless the critically limited global emissions budget was being divided fairly between it and the North?

Which is, we think, why a strategy that promises, at best, the begrudging support of the South will simply not do. Curves like this one are becoming more and more well known, and as they do, incrementalism is becoming manifestly implausible as a means of making it to a soft-landing corridor. The bottom line is that we have to start *talking* about the long-term curves, and the carbon budget that they imply, and *explaining* how we plan to balance that budget.

Tom Athanasiou and Paul Baer

CHAPTER 4

JUSTICE AND DEVELOPMENT

At the end of the last chapter, amidst a discussion of emissions budgets and possible futures, we showed you a terrifying, though quite plausible, "business-as-usual" scenario in which global emissions rise above the 450-ppm path almost immediately and *Southern emissions alone* crack the 450 ceiling in about twenty years. Here, we're going to take you down one last numerical byway and show you the relationships between per capita emissions, per capita energy use, and per capita income. As you may be getting a bit tired of graphs and numbers, here's why we're going to do it: *We intend to show that the South has excellent reasons to fear that limits on its emissions will, in time, become limits on its income and its development.*

Income doesn't equal welfare; this is now obvious, to the point that even mainstream critics of development insist that an overemphasis on GDP can be extremely misleading in evaluating human well-being. But to the extent that income is at least a proxy for well-being, and that we agree to talk about "nations" rather than classes within nations, it's income *per capita*—the average, not the total—that matters. India's GDP is larger than that of the Netherlands, but one country is rich and the other poor, and it's the per capita numbers that show you why.

And to the extent that emissions correlate with energy use, and energy use correlates with income, it's again the per capita numbers, not the totals, which matter. India uses six times as much energy as the Netherlands and emits six times the CO_2, but if you average across national populations, you see why India is a developing country and the Netherlands isn't: Both the Netherlands' per capita energy use and its per capita emissions are ten times as high as India's.

All this, of course, is patently obvious, and it may seem bizarre that we belabor it. But sometimes the obvious facts are the crucial ones. Thus, we need to restrict global emissions—drastically and soon—and this means that we need a global climate accord. And this, in turn, means that we have to find a fair way to divide up a finite "atmospheric space." To be blunt: Unless you think that global apartheid is a realistic and desirable option, the issue is, necessarily, fairness in a finite world.[1] And at the end of the day, there just isn't any way to conceive of such fairness except in per capita terms.[2]

A PER CAPITA LOOK AT THE BUSINESS-AS-USUAL WORLD

Look, now, in figure 8, at per capita emissions in the A1 "balanced" scenario that we presented in the last chapter. Note that average Southern emissions start at less than 1

Tom Athanasiou and Paul Baer

FIGURE 8: Projected Southern and Northern per capita CO_2 emissions pathways for IPCC's A1B scenario, compared to per capita pathway for stabilization at 450 ppm. Data from the IPCC's *Special Report on Emissions Scenarios*, 1999.

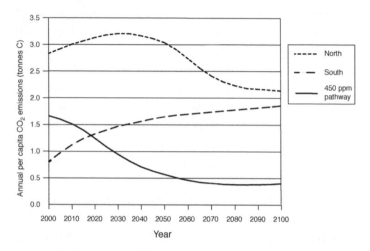

metric ton of carbon per year in 2000 and rise to just over 1.5 tons in 2040, while Northern emissions rise to over 3 tons per person per year in 2030 before dropping gradually back to just over 2 tons. Again, keep in mind that this is a "nonintervention" scenario; it assumes no climate policy whatsoever, not even the Kyoto Protocol. Any reductions in emissions are made either for strictly economic reasons (e.g., cleaner fuels are cheaper, or efficiency is increasing) or for other environmental reasons (e.g., as a side effect of reducing acid rain or "traditional" air pollutants such as particulate emissions).

This kind of analysis passes rather lightly over differences within the South. Singapore is still a "developing country," at least as far as the Kyoto Protocol is concerned, but it hardly belongs in the same category as, say, Sierra Leone. Nevertheless, this is a back-of-the-envelope analysis (albeit one done with sophisticated computer models), so bear with us, and note the third curve, shown here as a solid line. This is our old friend the 450-ppm path, the one that will lead to a warming of *only* 2°C. If we're lucky.

And as you can see, it describes a far more demanding path.

This "balanced" kind of North-South emissions convergence just isn't going to work, not even given A1B's rather forgiving population assumptions.[3] For while the A1B future would arguably be fair if we had enough atmospheric space left to take it, we don't. In fact, the 450-ppm soft-landing path, shown here in per capita terms, falls below projected Southern levels in less than two decades, before dropping later in the century to less than 0.5 metric tons per person! Even if the North magically disappeared, A1B-style Southern emissions growth alone would rapidly blow the global carbon budget.

The point in all this is hardly to blame the South, or, indeed, to make a big deal about population, which is

Tom Athanasiou and Paul Baer

emphatically not the problem here. The point, rather, is that a per capita analysis allows us to quickly *prove* a crucial point: The South can't develop the way the North has. It's just not possible.[4]

To round out this argument, we need to give you one last indicator, *energy intensity*, which is the ratio of energy use and economic activity.[5] And we need to talk about the relationship between greenhouse gas emissions, energy use, land use, and development.

To state the obvious: No one actually *wants* to emit greenhouse gases. Our carbon emissions, for example, are byproducts of other activities. By burning fossil fuels, we power planes, trains, and automobiles, keep ourselves warm or cold; we smelt ore to make metal, cut trees for building and manufacturing, clear and plow land to grow crops and raise cattle. Given the nature of our civilization and our sense of wealth, there's little doubt that, on average, the better off we are, the more we undertake such emissions-producing activities. This is, very crudely, the economic meaning of "development": more of the goods and services we want, and more energy used and greenhouse pollution created to produce those goods and services.

It won't give away the plot to say that development must soon become "sustainable development," that goods and services must be "dematerialized" and "decarbonized" (by using fewer non-renewable resources and emitting less CO_2)

to the point where neither their production nor their consumption would destabilize the ecosystem. This is all well known, and all to the good. But it's not the whole story. For just now, even as we're trying desperately to find some traction on the road to sustainability, it turns out that Southern countries—even though they're more vulnerable to climate change—are hesitant to agree to limit their emissions. Our point is that they're absolutely right to hesitate, and that it's long past time for Northern environmentalists to understand why.

It's not simply that the Kyoto Protocol has been so weakened that it will barely reduce rich-world emissions. Nor is it only that for decades, but especially since the fall of the Berlin Wall, the world has been beset by a breakneck globalization in which Europe and the United States have jointly pursued policies harshly inimical to Southern development. It's also that there's a long-standing lockstep between carbon emissions and wealth, and that the developing countries quite justifiably fear—all hope about sustainable development aside—that emissions limitations will permanently condemn them to a level of material well-being far lower than that enjoyed by the rich world.

Recall that we've been using three metrics of development: per capita income, per capita energy use, and per capita emissions. While there are limits to such simple measures—the most significant being that per capita emissions

Tom Athanasiou and Paul Baer

would become entirely irrelevant in a solar/hydrogen economy—it's also true that they're *not* irrelevant today, that industrialization generally means that the per capita demand for energy services increases more quickly than per capita income, and that energy intensity thus *increases before it begins to decline*.

There have been exceptions to this pattern. For example, energy intensity in China is steadily decreasing as it uses coal more efficiently and thus reduces its traditional (non-greenhouse) air pollution. But even here, China's rapid economic growth has meant that, despite its dropping energy intensity, its total energy use and its total emissions have both continued to rise. Development, in other words, has always meant rising emissions, and while the link between them has weakened, we shouldn't imagine that it's going to be easy to break.

One final thought experiment: Currently, the South's average energy intensity is about three times higher than the North's, and, clearly, this has to change. So imagine that the South enjoyed a magical efficiency revolution that instantaneously reduced its energy intensity by a factor of three. Its energy use, however, would stay the same (remember, this is magic), so its GDP would immediately triple, and thus there would be real progress toward developmental justice. But would it solve the problem of North-South inequality?

No. The South uses energy at about a fifth of the per capita rate of the North, and our magical decrease in Southern energy intensity would do nothing to change this. The South would still enjoy only one-fifth the per capita income of the North. To redress this remaining inequality, the South would have to get the same per capita energy budget as the North—as well as the same energy intensity—and until energy decarbonization becomes a near total reality, this relationship between per capita emissions, per capita energy use, and per capita income will continue to hold. In the meantime, if Southern energy use rose to the North's current level, and if that energy use came from the same mix of fuels that the North uses today, Southern per capita emissions would also, and by definition, reach today's Northern levels. It would be equality, but it would be catastrophe as well.

Thus, if the South is to become as wealthy as the North is today (to say nothing of ever catching up with the North, which plans to have moved on to new heights of wealth and consumption), it will have to do so with *much* better technology than that now used in the North. And even if Southern negotiators accept that income equality is nowhere in the cards, the fact is that they're not particularly disposed to experiments that, they fear, will close off the only routes to progress they've ever had.

Actually, it's even worse than that. Imagine that everyone on this benighted planet was suddenly to have an equal per capita emissions budget, and that this budget was set low enough to keep us on our less-than-fully-precautionary 450-ppm path. What would happen? Well, the South would find itself with a budget about a tenth of the size of the North's current level. And while the North would have the same small per capita budget, it would nevertheless still be wealthy, enjoying as it does all the benefits of massive capital stocks built up over centuries of open-frontiers development, and it would continue to enjoy all of the options that wealth affords. It would buy more atmospheric space from the South, which—all else being equal—would be left to develop, if it could, with a tightly limited environmental budget. And the South would quite likely fail.

JUSTICE AS REALISM

We think there's a way out of this, and we'll explain why. But note first, if you please, that this is a story about money. Much as we'd like to see the atmosphere as, say, a celestial sphere, or as the Earth's precious, protective membrane, we have to face the reality that it's also an economic resource, one to which we all have inherently easy access. And we also know that only some of us have been able to use this resource to become powerful and wealthy. This, of course, was in the past, and the past is unalterable, but the same cannot be said about the future. Which presents us, finally, with a crucial moral and political question: Why, now that the atmosphere's ability to absorb carbon is manifestly and critically scarce, should people in some countries continue to enjoy more of it, and thus more wealth, than others?

The answer, of course, is that there are not and cannot be any such reasons, save perhaps for raw power. And this, perhaps more than anything else, provides our rationale. For whatever one's position on "the commodification of nature" or the perils of tradable pollution permits, this remains: The "right" to dump greenhouse gases into the atmosphere has significant economic value, and the economic advantages of this right cannot be ignored.

The issue here is distributive justice. But understand that in a world beset by ecological crisis, distributive justice must mean more than it did in the past. It must include not only the fair distribution of wealth, resources, and opportunities, but the fair distribution of "impacts" as well. Because the elemental truth is that as the storms become more violent and the droughts more fierce, some of us will be hurt far, far more, and far earlier, than others. The rich will be able to hide, but the poor will not, and neither will the plants and the beasts. And because the "ecosystems people"—indigenous peoples, farmers, fishermen—who rely directly on nature for their livelihoods are the most threatened of us all, the real priority is finding a politics that makes a low concentration cap possible; it's at least as crucial as finding a fair distribution of emissions rights.

Tom Athanasiou and Paul Baer

It is, as Tom Paine said, "the good fortune of some to live distant from the scene of sorry," and this is perhaps even truer today than it was in his day. High-flown schemes will be of little use to the victims of tomorrow's killer hurricanes, unless these schemes are tied to and enable real change and honest hope. Climate change must be minimized, but at this point severe impacts are entirely inevitable. The harm these impacts bring to the poor— always the most vulnerable—must be minimized, and then alleviated, while the "burdens" of "adapting" to climate change must be honestly addressed, fairly distributed, and adequately funded. Anything else would be unjust and lead inevitably to distrust, bitterness, and failure.

This isn't going to be easy, and in practice the whole sustainable development agenda is going to have to be dragged into the picture. Leave out the details for a moment and note only this: A just adaptation program will be neither simple nor cheap. Inner-city residents in Chicago are already dying from abnormal heat waves, even as South Asian peasants are suffering floods more terrifying than any in living memory. Disaster relief as we know it will not do, and the questions, here as everywhere, come down to democracy and money. One question, in particular, is key: Who pays?[6]

The answer must be that the rich must pay, and not only because they're responsible for the problem, though they are: In 1990, the industrialized countries were responsible for 75 percent of all CO_2 emissions, as well as 79 percent of the CO_2 still in the air and 88 percent of the human-caused warming.[7] They must pay as well because their riches trace back to a past in which the world was open. And they must pay because only they can, and because if they don't, the warming will quite certainly prove unstoppable.

It has become common to justify the need for equity in narrowly realist terms, to argue that a certain measure of equity is a matter of national or economic self-interest or even, at the extreme, of national security. Without sufficient equity, so the argument goes, there will be political and social upheaval, which will eventually "blow back" to harm us, even in our most secure enclaves. The case for equity, in other words, is often reduced to the case for its instrumental utility as a precondition of both peace (or at least stability) and environmental sustainability.

We grant the point, but only to a point. We are, like so many others, struggling toward a new geoecological realism. But we also believe that equity—defined specifically as equal rights to global common resources—must be affirmed as a foundational ethical and political principle, a basic element of the unfulfilled enlightenment project of human emancipation. And here's the twist: We believe that this "moral vision" is itself part of the game, and a powerful political force. The way forward, in our view, lies in taking "the just" and "the realistic" as two sides of one single over-

Tom Athanasiou and Paul Baer

arching imperative, the two sides of the integrated vision we so desperately need.

The large countries of the South, and China and India in particular, do not require U.S. permission to burn their coal. Nor, despite their vulnerability, must they agree to a climate treaty that they regard as patently unfair, particularly if its unfairness lies in foreclosing their "development." Southern leaders are, moreover, entirely aware that the North owes them a huge ecological debt, that it rose to wealth and power in part though the high-carbon history that is now causing the climate crisis. Given this, it's quite unreasonable—and entirely futile—to suggest that the South accept restrictions on its development that perpetuate the disproportionate pollution of the North. It's just not going to happen.

Or put it this way: A climate treaty that indefinitely restricts a Chinese (or Indian) to lower emissions than an American (or European) will not be accepted as fair and, finally, will not be accepted at all. Climate equity, far from being a "preference," is essential to ecological sustainability. At the end of the day, it's just that simple.

CHAPTER 5

A PER CAPITA CLIMATE ACCORD

To review: We need a precautionary global emissions cap, one that translates an *impacts target* into a temperature target, a *temperature target* to an emissions pathway, an *emissions pathway* to a carbon budget, and a *carbon budget* to a *global emissions allocation accord*. And it isn't going to be easy to define such an accord, let alone bring it into force, for even a dangerously high temperature target—a maximum average warming of 2˚C—demands an extremely rapid drop in global emissions, to a level well below the current world average and far, far below the utterly unsustainable emissions levels of the North.

Obviously, the North isn't looking forward to such cuts. And the United States, in particular, is going to put them off as long as it possibly can. In this context, it's clear that the next step in the climate regime, the one that comes after Kyoto's entry into force, is going to have to be significant and fair, and it's going to have to begin bending down the curve of developing country emissions. Don't misunderstand us: the North must take the lead. But once it does, the South is going to have to follow. One way or another, the developing countries are going to have to accept emissions caps.

The question is how.

For years now, the South has argued that it won't accept emissions caps until the North *has already acted* to cut its emissions. Southern diplomats and activists speak repeatedly of their "right to development," and, as we've tried to show, such a right must necessarily involve a huge increase in emissions, a large-scale renewables revolution, or both. And even if we manage a massive and thoroughgoing clean energy revolution, there's still a long-term reckoning in the cards: If "development" is to eventually mean equality of income, then there will have to be equality of emissions rights as well.

It's a fine mess, and it's time, now, to talk about the terms of the climate accord needed to cut a path out of it. And note, please, that while the problem is fantastically complicated, the solution has to be, if not simple, at least simple enough to explain, and to justify as fair, all around the world.

Here are the essential features of such a climate accord:

➤ It must be adequate to the science and be defined by the impacts that we decide, together, we're willing (and able) to tolerate. It must frame the problem in precautionary terms and offer at least a chance of capping global temperature at a tolerable level. If a proposed accord doesn't do so, it should not and will not be taken seriously.

➤ It must be global. The North and the South must both sign on, though not, of course, on the same terms. And there must be strong penalties for cheating and noncompliance, which, crucially, means that global trade rules will have to change to allow climate-cooperating states to use carbon tariffs and border taxes to isolate free riders.

➤ It must drive a rapid process of technology "leapfrogging" in the South. Indeed, it must *prevent* the South from building a fossil fuel infrastructure that mimics the unsustainable path of the North, even as it drives decarbonization in the North.[1] The key is that the South's leapfrogging must be part of a genuine development process, and the North must pay for it. This can't be just another false promise.

➤It must be based on equal per capita emissions rights, *which would be phased in over time.* Any other framework would enshrine a system in which some people have greater access to the atmospheric commons than others. This would not only be morally intolerable, it would also fatally undermine the global cooperation we need to fight climate change.

➤ It must take proper account of different national circumstances. Some countries have far greater resource endowments than others, and some are far colder. And some, of course, are far richer. All of this must somehow be taken into account, in a systematic way that does not endanger the overall per capita framework.

Tom Athanasiou and Paul Baer

➤ It must be practical and efficient. Countries with unused emissions allocations must be able to transfer them to over-budget countries and get money in return. This probably means "emissions trading," though not a simple kind of trading in a simple kind of market (see chapter 6). The "financial mechanisms," however they turn out, must be honest, well regulated, comprehensible, and fair. And the money countries get by selling their allocations must be verifiably earmarked for clean energy and sustainable development.

➤ It must begin with a trust-building phase, and this phase must start to rapidly bend down the emissions curves. Note, in this regard, that trust building has always been one of the principal goals of the Kyoto Protocol. We have to hope that, even after all the hits that Kyoto has taken, it's still good enough to clear the way for a global accord.

➤ It must go into effect in 2012, when Kyoto's first commitment period expires. Nothing else would have a chance of being adequate.

It's quite a list, but in practice it all boils down to something pretty simple: There has to be a phased transition to a system based on *rapidly shrinking* per capita shares, and countries must be able to sell their unused shares and use the money for clean energy and sustainable development.

In practice, of course, this will get complicated—everything related to the climate gets complicated in practice. But a per capita accord like this one might actually work, and as far as we know, there's no other idea out there whose proponents can convincingly make the same claim.

THE PER CAPITA ACCORD IN PRACTICE

There are, of course, lots of questions. How are we going to democratically determine what level of impacts—and thus what temperature target—is acceptable? How can we take different national circumstances into account without being trapped by a nightmare in which every country pleads that its "unique" circumstances should allow it to exceed its allowance? How far should developing countries be able to exceed their per capita shares in order to develop, and for how long? Where would the atmospheric space for such developmental overshoot come from? And how should we deal with the Malthusians, who damn per capita allocations as a "breeder's charter" that would encourage population growth around the world?

Some of these are tough questions, but even tough questions can be answered within a context of increasing global democracy and cooperation, and without such cooperation, there's little hope in any case. Furthermore, global cooperation is just what the per capita climate accord is all about, the kind of genuine cooperation that can only solidify with-

Tom Athanasiou and Paul Baer

in a regime designed to recognize real interests, even as it helps to refashion those interests into shapes appropriate in a fair and finite world.

Moreover, this proposal is as practical as any adequate plan can possibly be. The North's obligations to the South are many, but we know of no better way to translate the North's obligations into effective action than by means of a well-designed per capita climate accord.

Under such a system, most poor Southern countries (remember that as far as Kyoto is concerned, rich countries such as Singapore and Saudi Arabia are still part of the South) would have surplus permits to sell, while the Northern countries (as well as the Singaporeans and Saudis of the world) would be net buyers. The resulting cash flows—not aid but *payments* for the use of someone else's environmental space—would then be used to finance a transition to low-carbon energy in the (poor) South, while at the same time motivating the industrialized world to transition to clean energy. The North, after all, would find itself over-budget and would have to either decarbonize its economies or continue to pay, indefinitely, for its excess pollution.

We don't pretend that it would be easy to manage such a system, or to ensure that the payments are indeed used for energy leapfrogging and sustainable development, though we do have some ideas. The main point, though, is that the more rapidly the global emissions budget fell (a global political decision), and the more rapidly we cut over to a per capita allocations framework, the more money the North would have to pay to purchase atmospheric space from the South, and the more the South would see in its leapfrogging accounts. The system, in other words, would be elegant, fair, and simple. Simple but, we think, not too simple.

Would it, however, be winnable? Clearly, this is an open and urgent question. But if you're tempted, at this point, to conclude that a per capita accord is not in the cards, stop for a moment to examine your objections. Are they to a per capita accord, per se, or do you rather think that *no* adequate climate deal is possible, not, in any case, in time? We ask because, frankly, we've found that most objections to per capita, if carefully probed, devolve into a more generalized pessimism about our chances for stabilizing the climate, to the sense that regardless of what we do, our futures are already written.

Our view is that it's time for honesty rather than despair. Like this: Attempts to base a global climate accord on any ground other than equal per capita emissions allocations are doomed to fail, for they would continue the historical subsidy of the underemitters to the overemitters—of the poor to the rich—that is such a crucial, and unacknowledged, pillar of this civilization. The historical and continuing overuse of the atmosphere by the North can be seen to be the accrual of a massive ecological debt, and this accrual must stop before a workable climate treaty can be put into place. And

Tom Athanasiou and Paul Baer

as far as practicality is concerned, the real question is if, after Kyoto's entry into force, Southern negotiators will be willing to step forward to support a per capita accord. If they do (and they may) we'll be in a whole new ball game.

Be clear: *This is a brief for equal emissions rights, not equal emissions.* Remember that neither people nor countries actually want to emit CO_2 or other greenhouse gases; what they want is energy services and increasing well-being, and, insofar as they think it necessary for increased well-being, they want economic growth. Remember that burning fossil fuels comes at a high price in local air pollution, and thus in terms of asthma, cancer, and all the rest of it, and remember the environmental destruction associated with their extraction. Remember that few Southern countries have huge fossil resources, and that fossil dependency generally means high and erratic demands on foreign exchange budgets. For all these reasons and many others, the South need not see climate justice as the justice of following the North down the fossil-fuel path.

What the South wants is development, which is why the climate accord is such a fantastic opportunity. A truly fair accord, were it placed on the table, would find Southern friends in profusion, for it would offer more than cynical rhetoric and development as usual. It would offer, instead, a means to make "sustainable development" into something more than a cruel, false promise.

It's not an opportunity we get every day.

"CONTRACTION AND CONVERGENCE"

The idea here is not ours. The merits and demerits of a climate treaty based on tradable per capita emissions allocations have been discussed in academic, activist, and policy circles for more than a decade, though it was *Global Warming in an Unequal World*, published in 1991 by the late Anil Argawal and Sunita Narain of New Delhi's Centre for Science and Environment (CSE), that put the core idea—equal per capita rights to the atmospheric commons—into political motion.[2]

The best-known articulation of the idea is "contraction and convergence," which Aubrey Meyer, director of London's Global Commons Institute, has been tirelessly promoting for many years.[3] The term "contraction" refers to a reduction of global emissions from today's unsustainable levels to future "safe" levels, while "convergence" implies that at the same time, developing country emissions allocations would be allowed to increase in the interests of development, while rich-world allocations would drop. The result of these transitions would be a global convergence to equal, and low, per capita allotments.

The contraction-and-convergence framework assumes that convergence takes place over some transition period (by, say, 2030) and that allocations are tradable, so that per capita emissions themselves may or may not actually converge.

Tom Athanasiou and Paul Baer

This is a key point, so note that it's not some sort of rich-world trick, and that, for example, India's Centre for Science and Environment takes the same position. The goal is convergence of emissions *rights*, and decarbonization of energy systems, not convergence of emissions themselves.

Beyond particular schemes, key Southern voices have long insisted on rights-based (per capita) allocations. Examples are many, but the declaration of the 1998 meeting of the Non-aligned Movement can perhaps stand for them all:

> Emissions trading for implementation of (GHG reduction/limitation) commitments can only commence after issues relating to the principles, modalities etc., of such trading, including the initial allocations of emissions entitlements on an equitable basis to all countries, has been agreed upon by the Parties to the Framework Convention on Climate Change.[4]

Also, it's worth noting that Dr. R. K. Pachauri, the new chairman of the IPCC, is the director of the Indian group TERI, which called, in early 2002, for climate action

> through comprehensive international participation and agreement on the final level at which to stabilize the concentrations of GHGs and on medium-term targets for reducing emissions. Carbon trading

arrangements based on an equitable per capita allocation also need to be adopted.[5]

The idea, in other words, is pervasive, though not, so far, within the climate negotiations themselves. An increasing number of organizations and politicos, including a bloc of European environment ministers, a variety of international environmental nongovernmental organizations (NGOs) as well as traditional NGOs such as the Red Cross and Christian Aid, Britain's influential Royal Commission on Environmental Pollution, the former co-chair of the IPCC's Working Group One, and a rich variety of Southern politicians, have explicitly endorsed it, and many others have adopted the per capita framework, though not the term "contraction and convergence." Further, both India and China have repeatedly signaled (or so we've heard, for these things are rarely written down) that when the time comes for them to accept emissions targets, nothing but per capita allocations will even be considered. The terms by which allocations are defined must, as a Chinese delegate to the climate negotiations once insisted, be "rational."

We agree. For, from the point of view of both basic ethics and enlightenment philosophy, the case for equal per capita rights is an obvious one. Yet, at the same time, human rights are under siege around the world, and this proposal implies a radical expansion of such rights, one that actually expands

　　　　　　　　　　Tom Athanasiou and Paul Baer

them into the new territory of *economic* rights to global environmental resources. Why, then, do we imagine that the idea will find political traction in the "real world"?

The easy answer is that, as the references to India and China imply, nothing else will yield a global climate accord. A historic choice will be made during the next decade, as the next phase of the climate treaty is thrashed out, and appeals to "realism" and incremental decision making do nothing to alter this rather brute fact. Explosive as the per capita issue is, we do not believe that it can be finessed.

That's the easy answer, but there's more. Here are three key points:

PER CAPITA WOULD BE EFFICIENT

This is a capitalist society, and whatever else this may imply, it certainly means that the "costs issue" is unavoidable. So note that a per capita system could well be less expensive than the alternatives. Jan Pronk, the Dutch environmental minister who served as midwife to the historic Bonn Compromise, himself averred, in an interview just before 2000's Hague climate conference, that tradable per capita allocations would be the most logical and efficient approach, as it would quickly create a large carbon market. Others have reached the same conclusion, and why not? This is just basic neoclassical economics, isn't it?

What matters, in economic theory, is that emissions rights

be priced, and that "the market" be as large and efficient as possible. This could come down as Kyoto-style carbon trading or a global auction, and we may well need a global development fund to make it work. But trading is very much on the table. And while trading is distasteful to many in the global-justice movement, there's plenty of evidence that well-designed tradable permit systems (and, by the way, well-designed trading systems are never *pure* trading systems; they always need strong regulatory frameworks) are sensible ways to reduce pollution. And it's easy to see why—trading gives those who can cheaply reduce their pollution a strong incentive to do so, because they can then sell their unused emissions allocations to others. Economic theory says that, to a greater or lesser extent, depending on the details, trading will achieve a given reduction in pollution at the lowest total costs, and, in fact, it often works out that way.[6]

There are, of course, serious problems with trading. For a discussion of all this, see chapter 6.

PER CAPITA WOULD BE FAST

With regard to the need for speed and early action, the Dutch National Institute for Public Health and the Environment (known by its Dutch acronym, RIVM) recently published a careful comparison of a sophisticated "phased" proposal (one that avoids the grand strokes of the per capita deal) and a contraction-and-convergence approach.[7] It's an

Tom Athanasiou and Paul Baer

honest bit of work that sets out the phased alternative in terms that are as compelling as possible, and as fair. The basic idea is that when a country "graduates" to some key threshold, one defined in terms of both wealth and emissions, it would have to take on emissions limitation commitments.

The conclusion was even more interesting: Plausible "graduation thresholds" on wealth and emissions do not bring enough Southern countries under the cap in time to hit 450 ppm, and contraction and convergence, in comparison, would be faster.

PER CAPITA WOULD BE GLOBAL

Many of the advantages of the per capita approach follow from its global nature, and any other global system would share them. However, as it's unlikely that a workable global regime can be built upon any other foundation, this is a pretty academic point—in practice, the advantages of the global approach are probably the advantages of per capita alone. For example:

➤ Per capita would solve the "leakage" problem—which is a technical way of saying that capital flight wouldn't be a problem because there wouldn't be anywhere to go. Nonglobal regimes, on the other hand, necessarily suffer from leakage, because if one country has accepted an emissions cap and another hasn't, then, all else being equal (and of course it's not), the first is at a competitive disadvantage. The result is that national industries are hurt, trade wars are threatened, and, in general, leakage problems undermine support for climate protection.

In a global regime, the logic is quite different. Everyone would be in, save for the "free riders" that seek to stand outside the regime, reaping its benefits (a safer climate) while doing none of the work. Thus, the parties to the accord would have to protect themselves from the free rider's exports, perhaps with border taxes, and these might, in fact, be key to the accord's enforcement. But these protections would themselves be global, not ad hoc; and far from threatening trade wars, they would demand the reform of the trade system itself. A big plus, this.

➤ Per capita would solve the "baselines" problem that bedevils the Kyoto system. This is an important point because, absent a global accord, greenhouse politics will remain dominated by "project-based" approaches like Kyoto's Clean Development Mechanism (CDM), wherein Northern countries can get one-off credits by financing (hopefully) virtuous development projects in the South.

There are lots of problems with the CDM, but let's assume it works the way it's supposed to, and that when a Southern country agrees to host a project, it gets the benefits, say a modern gas-fired power plant that it would not otherwise be able to afford. Nevertheless, there's also a downside to accepting such projects. As things stand today, absent a global accord, developing countries must assume that they'll someday be forced to accept emissions caps on unfair terms. These terms would be relative not to their population, but rather to their historical emissions levels, which would be taken as their emissions "baseline" and used to calculate their future share of the atmospheric space. What this means in practice is that absent a fair global deal, developing countries have a perverse incentive to keep their emissions high, in case they're forced to accept baseline-related caps. It's a version, a suitably modern and abstract version, of the so-called tragedy of the commons.

It's hard to sum all this up, but here's an attempt: Only a rights-based framework can honestly offer the South a fair share of the atmospheric space. And only a rights-based framework can ratchet up the North's willingness to pay to the point where it's even plausibly consistent with the scale of the needed international investments (or, for that matter, the scale of the North's ecological debt). Without such a framework, any climate-related North-to-South payments would be marked as "aid," and we know where that leads us.

The way out is a system in which atmospheric space is allocated fairly, and then "belongs" to those to whom it is allocated. The way out, in other words, is not the "burden sharing" that the climate people are always talking about, but rather a system of "resource sharing" based on equal rights to the atmospheric commons.

PROBLEMS, ISSUES, OPEN QUESTIONS

As noted above, the strengths and weaknesses of the per capita idea have been debated for some time. There's still a great deal of work to do, but some of the Big Questions, in fact, already have good provisional answers.

THE POPULATION QUESTION

Ever since the late 1980s, when the per capita allocation of carbon-emissions rights was first proposed, Malthusians both green and otherwise have attacked it as a "breeder's charter." This is actually quite mad—as if developing nations, with all the challenges they confront, would encourage high fertility rates to increase their carbon emissions allocations! But as rich-world populationists are still influential, the friends of per capita typically move to disarm this charge by including a "population cutoff year" into their projections. There are a variety of other proposed solutions to this alleged incentive problem, but to be frank, it's the least of the obstacles.

WHO GETS THE MONEY?

Another common argument against a per capita climate

accord is that the resulting financial transfers would not in fact be used for clean energy development, but would instead be used for monumental, World Bank–style development projects, or even wind up in Swiss bank accounts.

This is a real issue, for it raises the problem of "green conditionalities" and, once again, the North-South double standard. Why, after all, is it always corruption in the South that's the *structural* problem, while corruption in Washington, or in the suites of the transnationals, is just a matter of a few bad apples? And certainly we do not judge the democratic nature of the Saudi royal family before we pay for oil; why should we, then, do so for emissions rights?

These are good questions, but the conditionalities here are nevertheless necessary. The immediate point, after all, is clean-energy leapfrogging in the South, and this means that there must be controls on the uses of any payments. And as a matter of political reality, support for the per capita accord will be far easier to build if balky politicians can be assured that the money will be spent as is intended.

We'll get back to all this in the next chapter, but note, for now, that the problems here are not fundamentally problems of sovereignty, but, rather, problems of democracy. There will have to be controls on the cash flows associated with emissions allocations, but the South, and indeed impacted communities within the South, must be equal partners in the creation and enforcement of these controls. That's the bottom line.

A practical per capita accord would differ from classical contraction and convergence in taking national circumstances into account. If it didn't, it would create its own forms of unfairness, since countries vary in their natural resource endowments, their climatic conditions, and their technical and social capacities, and thus "straight" per capita allocations would lead to vastly different costs of energy in different countries.

But the rules by which such circumstances would be creditable must be carefully controlled, to prevent the whole allocations system from collapsing under the weight of side deals and special pleading. (One wag anticipated "some fabulous formula based on land area, climate variation, fossil fuel, hydro resources, and olive supplies that will make sense only to Werner Heisenberg.") This, fortunately, should be possible, particularly if deviations from per capita were approached by way of the principle of equity itself. The idea would be to distinguish between circumstances that, taken into account, would lead to a fairer deal—one with more equal outcomes— and other circumstances, say special pleading by oil-exporting countries, that would actually increase inequality.

The key distinction here is between short-term fairness, or *transition equity*, and long-term fairness, in which the per capita principle should only be modified on the basis of resource endowments and climatic conditions. South Africa is a good example of a country with genuine transition issues, since its emissions are already so high as to put it above any sustainable per capita budget, but most of its people are still extremely poor, in part because the infrastructure is so inefficient. This needs to be taken into account, but it can't be taken into account forever. Transition issues are just that, and transitions end.

HOW MUCH WOULD THIS COST THE NORTH?

Finally, there's the ultimate argument against per capita emissions allocations: that the industrialized countries, and the United States in particular, would simply never agree because, right or wrong, their governments and their people are simply not willing to pay. It's a strong objection, and at the end of the day, there's no real reply. The per capita accord appears to be the fairest and the most efficient (read cheapest) global option open to us, and thus our best chance of holding global warming down to tolerable levels, but if the major countries, like, say, just hypothetically, the United States, will not agree to its terms, then that bridge will have to be crossed, in time, by other means.

In any case, what kind of money are we really talking about? The answer, unfortunately, is not available, for no good economic models have yet been done of a per capita accord. Better projections will be available in time, but for the moment, we'll have to make do with coarse estimates.

For example, in a tradable permit system with a global emissions cap set at today's very high level, an *immediate* transition to an equal per capita allocation (which no one is proposing; the talk here is of a phased transition) would result in the trading of roughly 2 billion metric tons of carbon permits each year. Recent economic studies have estimated that for a cap based on the Kyoto framework, permits might trade in a global market at $20 to $100 per ton,[8] but estimates range up to $200 per ton or more for more restrictive global caps. It's not, actually, a lot of money—each $25 per ton increase would translate into an increase of only about 7 cents in the price of a gallon of gas—but since the United States alone exceeds its equal per capita share by more than a billion tons, it's not entirely trivial either.

We can at least put a rough upper bound on the price, by noting that pessimistic, anti-Kyoto analysts have put the cost at a *few percent of global economic output per year*. This would be a lot of money; with global GDP estimated to break $100 trillion in thirty or forty years, it would be trillions of dollars. But—and it's a big but—these numbers are not only cooked, they're also presented in profoundly misleading ways.

For one thing, the models behind them can easily be shown to be based on absurd worst-case scenarios in which all the myriad well-known opportunities to reduce emissions and energy use while saving money—including such

Tom Athanasiou and Paul Baer

obvious hardware-store banalities as compact fluorescent lights—are simply assumed not to exist.[9] And however much emissions reduction would cost, it still only comes to a tiny percentage decrease in the rate of economic growth. Think about it this way: The IPCC's Third Assessment Report includes the very mainstream economic assessment that stabilizing atmospheric carbon dioxide at twice preindustrial concentrations by 2100 would cost between $1 trillion and $8 trillion. It sounds like a lot of money, but compare it to world economic growth as predicted by these same economists (several percent a year), and it becomes all but invisible. At such a rate, the world as a whole will be ten times as rich by 2100, and people on average will be five times as well off. And according to climatologist Stephen Schneider and energy economist Christian Azar, two highly respected analysts, *adding the costs of tackling warming, even if they come to as much as 5 percent of global income (an implausible but typical estimate), would postpone this target by a mere two years*, from 2100 to 2102.[10] Similarly, meeting the terms of the Kyoto Protocol would mean industrialized countries "get 20 percent richer by June 2010 rather than in January 2010."

We'd like to think that most people, asked bluntly if they'd accept such a sacrifice in order to preserve the stability of the Earth and its climate for their grandchildren, wouldn't waste a lot of time agonizing about the decision.

CHAPTER 6

TRADING, TAXES, AND FUNDS

Warning: We believe that without its emissions trading provisions, the Kyoto Protocol would already be dead. And because we believe the way forward is by building upon Kyoto, we see the challenge of climate protection as inseparable from the challenge of making emissions trading, as we know it, into something fair.

Alternatively, there are other ideas to pursue: development funds and taxes chief among them.

Dead Heat argues for atmospheric resource sharing by way of rights-based, "one person, one share" emissions entitlements. But what, really, does such a proposal come to? Does it have any real chance of finding political traction? And is it a realistic approach to the grand North-South compromise we'll need if we're to avoid a catastrophic degree of climate change?

The last question is the key one, and its answer must finally turn on what Isaiah Berlin called "the sense of reality." Different people will answer differently, and none of us can say, just now, who is right and who is wrong. What we can do, however, is clearly explain to each other how we see the playing field, and what we see as the advantages and risks of different approaches. We're going to try to do that here, by reviewing the debate over international carbon trading and development funds.

Most climate-policy realists support trading, and for one reason above all others—they see trading as a means of maximizing global efficiency, and thus of maximizing our collective ability to respond to the threat of climate change. But suggest to them that the politics of "global common resources" and "environmental rights" are also essential, and they'll quickly balk. Professional environmentalists, in particular, will find a way to insist that it's not the job of the climate treaty to redistribute wealth. Press the point, and you'll be cautioned that the notion of "emissions rights," too publicly discussed, could destabilize the whole Kyoto framework.

The global-justice movement, for its part, has a different sense of reality. Propose *carbon trading as resource sharing* in global-justice circles, and you'll find that it's the resource sharing side of the equation that goes down well. Press the point, and argue that *some sort of trading* may be essential to any realistic way forward, and you'll be taken to task for timidity and self-delusion. In global-justice circles, carbon trading is generally seen as surrender to neoliberal ideology, and to market mechanisms that have been entirely captured by the corporate elites. Trading is only an invitation to the Enrons of the future, and a way for the rich to buy their way out of making real emissions reductions.

We, for our part, fear that this is a sterile debate, and that it threatens to drown out strategic rethinking. So let us begin again, this time with the Kyoto Protocol as our exemplar of "reality." And let's ask some tough questions: Doesn't our sympathy for emissions trading, and our patience for the whole disappointing Kyoto process, contradict our claims to speak for justice? Isn't the market itself an instrument of injustice? And speaking of Kyoto, isn't it doomed by the very compromises that saved it in Bonn and Marrakech? Isn't it so laden with "flexibility mechanisms" (like bad emissions trading and "sinks" rules) that it can no longer be counted as an honest first step, one taken by the North in recognition of its historic responsibility for climate change?

Such questions can clear the mind. And fortunately for us, and for our beleaguered sense of hope, most of our answers are no. The exception is the really tough question— Has the market itself become an instrument of injustice?— for here the answer is clearly yes. But even so, the implications for the trading debate are hardly obvious.

Begin here: Without emissions trading, the Kyoto Protocol would have died long ago, probably at Kyoto itself. This is the beginning, though not the end, of climate realism. Furthermore, without trading (or some similar market mechanism) the costs of cutting carbon would be too high (too *unnecessarily* high) for any climate treaty to have a hope of surviving, let alone working. Some sort of market mechanism

Tom Athanasiou and Paul Baer

is historically necessary, and this has to be stipulated right off the top, if only so we can move beyond elementary confusions to more interesting ones. The real question, as we see it, is how carbon markets can be made fair, and if, having managed such a Herculean task, we'd still recognize it as trading.

First, though, we need to review the familiar critiques of carbon trading and admit that some of them are a whole lot better than others. All of us, on all sides of this issue, are going to have to learn to tell the difference.

THE BAD, THE UGLY, AND THE GOOD

The bad critiques are many, but they may perhaps be represented by the beautifully printed "Carbon Credits" that, at the climate convention in November 2000, protesters from the Rising Tide network dropped from the occupied rafters of The Hague conference hall. They were good to see on that grim evening, though their message—that trading would necessarily usher in an age of "Carbon Colonialism"—wasn't a particularly cheery one. Not as, hot behind closed doors, the fight was raging to save Kyoto from stalemate and asphyxiation.

And yet that very fight, and the fact that even the protesters seemed to want Kyoto to come out of it alive, indicated that for all the clear simplicity of their argument, it wasn't really to be taken as anything more than theater. They decried the presence in the halls of groups such as the Emissions Trading Association, but, after all, it was Kyoto itself that was on the mat. Were they protesting for or against it? They may have known, but it sure wasn't clear.

Here, because they are instructive, are three arguments against trading, all taken from the British Rising Tide Web site:[1]

A bad one:

> "Carbon trading does not work. It has undermined the tiny reductions proposed in the Kyoto Protocol."

An ugly one:

> "The main argument against carbon trading is that it requires the privatization of the atmosphere. Property rights in this vast new speculative market will then be allocated by those who trade fastest and those who already pollute the most."

And a good one:

> "A global carbon market cannot be monitored or controlled, and the legal framework is unlikely to ever be strong enough to counter the huge incentives for cheating."

The "bad" one, as we've noted, is simply wrong. Without trading, there'd be no Kyoto to undermine. Without trading (and we'll grant that Kyoto's trading rules are pretty ugly), we wouldn't be awaiting Kyoto's ramp-up and speculating about the path forward to a global treaty. We'd be talking instead, in some despair, about picking up the pieces, and just now, in our post-9/11 world, this wouldn't be very easy. And even if we could find a way, it would require some sort of carbon market.

The "ugly" one, for its part, is also wrong, but in a more subtle way. Does trading require "the privatization of the atmosphere"? No, it does not. But it does require the use of markets, and because markets (which are, by the way, institutions that long predate capitalism) have become gears in a profoundly undemocratic economic machine, their use must inevitably risk confederation with elites, opportunists, and entrepreneurs of various unsavory varieties. But this, please note, is not simply a problem with trading. Any system in which the North helps to pay for clean energy development in the South would mean large public cash flows, and where there are public cash flows, there will be people trying to "privatize" them. Which is, by the way, another good reason for emissions entitlements to be treated as rights rather than being *grandfathered* over to current emitters, or handed out on an ad hoc basis. Rights are not, first of all, instrumental things, but they do have their uses. And as a practical matter, rights-based allocation would help prevent "property rights in this vast new speculative market" from coming, in the end, to "those who already pollute the most."

Which brings us to the "good" argument—that the "legal framework is unlikely to ever be strong enough to counter the huge incentives for cheating" inherent in these strange, new, almost fictitious international markets. That in a world dominated by the culture and instrumentalities of capitalism, emissions trading will be a tool in the hands of the rich. That they will insist that allocations be grandfathered and, finding sympathetic ears among the politicians, will lock out any other possibilities. That emissions markets can perhaps work within nations, but that global emissions markets will be impossible to police. That collusion and gaming will be the norm. That overlapping rules and opaque dealings will yield labyrinths of unaccountable transactions. That cartels will form, and price wars rage, and that even as the droughts spread and the death tolls surge, the powerful will do whatever they need to do to keep the price of carbon down.

It's a real and present danger, and if we succumb to it, we'll be in real trouble. But note, if you will, that Rising Tide is hardly alone in its fear that carbon markets will be Enronized. In fact, the fear that these markets—and indeed all global markets—are outrunning international legal authority is one shared by buttoned-down Washington policy wonks, a variety of European think tanks and ministries, and even some carbon traders themselves, the very sorts of folks that Rising Tide usually demonstrates against. And it's a damn good thing that it is, because a lot depends on getting this right: If the carbon markets come to resemble the California energy market, the Kyoto Protocol will quite predictably collapse, and the way things are going, it could take the whole "climate process" with it.

WHAT'S THE ALTERNATIVE?

The question is if we have any alternative to carbon trading, and many good people think we do: Ross Gelbspan, for example. Gelbspan, a well-known writer, is a tireless proponent of the basic truths of climate change: its impacts will be very severe indeed, and they will come soon; to avoid catastrophe we need a crash program of energy leapfrogging in the South; this will take *real money*. With equal forthrightness, he explains that in his opinion, trading will simply not work.

> Domestic cap-and-trade programs—like the U.S. trading program set up to reduce sulfur dioxide emissions—were relatively successful because they are easy to monitor and enforce.... At the international level, however, the system of "cap-and-trade" totally breaks down. It is not monitorable. It is not enforceable. Moreover, it is plagued by irreconcilable equity disputes between the countries of the North and South.[2]

Ross's alternative, the World Energy Modernization Plan, relies instead on three interacting strategies:

➤ "A change of energy subsidy policies in industrial countries."

➤ "The creation of a large fund to transfer renewable energy technologies to developing countries."

➤ "The replacement in the Kyoto framework of the ineffectual and inequitable mechanism of international emissions trading with a progressively more stringent Fossil Fuel Efficiency Standard."

There's a great deal to recommend this approach, though we, obviously, don't think that Kyoto's trading system is *doomed* to be "ineffectual and inequitable," or that, in the search for ways forward, emissions trading should be ruled out of consideration. And we, for our part, wonder if Gelbspan's plan for financing the energy transition—a "Tobin tax" on international financial transfers that he estimates would bring in $300 billion a year, or, if this doesn't pan out, an international carbon tax—has any real chance of being adopted.

A Tobin tax would be wonderful for all sorts of reasons—not the least of which is reducing the power of global capital—and the idea is extremely popular in the global-justice movement.[3] But, frankly, we don't think there's much chance that it will be in place in time to fund the greenhouse

Tom Athanasiou and Paul Baer

transition. A Tobin tax, after all, would be an *international tax*, and as such would be bitterly and tirelessly opposed by a broad transnational coalition of mainstream and lunatic elites. Besides, who'd collect the tax? And who would decide how it would be disbursed, and under what terms? These are tough questions under any circumstances, but they'd be even tougher with a tax that isn't directly related to energy or emissions. If Tobin was enacted, there would, and quite correctly, be demands on the money coming from all sides, and no guarantee at all that it would go for clean energy, or even for sustainable development. After all, where's the link?

A Tobin tax, of course, isn't the only possibility. An international development tax doesn't have to be based on capital flows, and a carbon tax might be a more practical and appropriate source of funds for a clear-energy transition. Certainly the idea has its friends, though, alas, we are not among them. It would still be an international tax, and, again, an international tax would pose an explicit challenge to the sovereignty of the national state and enflame the right wing. Not such a bad idea, perhaps, but do we want to pick this fight, too, and put ourselves in a position where we have to actually win it before getting back to the battle to stabilize the climate?

What if we put aside the idea of a tax and concentrate on the idea of a *development fund*? This idea, too, has been around for awhile; for example, there was the "Global Atmospheric Commons Fund," which the World Council of

Churches (WCC) was pushing at The Hague climate conference.[4] The idea here, explicitly based on the contraction and convergence model, is to collect funds from "countries which use the global atmospheric commons in excess of the convergence level," and then use these funds to "assist impoverished countries and those with economies in transition to move towards a non-carbon economy focusing on renewable energy sources such as solar, biomass, wind and small-scale hydroelectric." Meanwhile, the "wealthy polluting countries" would have to "pay the penalty for their excess emissions," which would act as "a strong economic incentive for them to initiate structural changes towards long-term technological and societal innovations in their economies."

Does this sound better than a trading system? The authors of the WCC proposal clearly think so, for two distinct reasons:

➤ It would be funded by user penalties, and "clearly, a user penalty differs significantly, ethically and legally, from a property right."

➤ "In contrast to the Kyoto Protocol's flexibility mechanisms, the industrialized countries would not be able to off-load their responsibilities by buying credits from other countries."

But, unfortunately, this is all pretty confused.

For one thing, this proposal does create a de facto property

Tom Athanasiou and Paul Baer

right, up to the level at which the penalty kicks in. For another, we'll all be a lot better off, in the battle for a just climate regime, if we stand for *equal rights* rather than "user penalties." To be sure, such rights must necessarily be construed as property rights, but these are rights to *common property*, and we see nothing at all disturbing, ethically or otherwise, about the notion of equal rights to global common property.

As to the second point: Despite the WCC's plea against allowing industrialized counties to "off-load their responsibilities," the net result of their proposal would be exactly this. Which is as it must be. The per capita proposal is *designed* to allow the rich to buy atmospheric space from the poor, and thus to not only provide the money needed for clean energy and sustainable development but also to reduce total global emissions reductions at the lowest cost. Besides, in the long term (and even in the middle term) the North will not be able to avoid substantial emissions reductions, *not if we intend to stay on a soft-landing corridor.* Because if we do, the per capita carbon budget will soon drop so low that Northern societies will be forced to rapidly decarbonize.

The core idea here, however, has powerful merits. For one thing, it would avoid the dangers of global carbon trading, which may, for all we know, forever elude regulation. And a fund-based system would offer clear opportunities for cash-flow control and monitoring. If we want the cash to end up being embodied as, say, solar arrays and small hydro systems rather than Swiss bank accounts, this is a pretty important consideration. And collecting payments into a global fund means that, at least potentially, they can be disbursed to local communities, regional civil assembles, and non-governmental associations within nations, as well as to national governments themselves. Such a fund would need a carefully designed, democratic, and just governance structure, but this is a problem that can be solved.

What's the downside? Collecting the money is the big one. However the details shook out, the right-wing would insist that this is only an international tax in disguise, and this would make it easier for them to kill the idea. And the problems of making a system like this both efficient and democratic would be anything but trivial. The global fund, after all, would be a hothouse of competing claims and competing ideologies, and would even pit Southern governments against local groups within the South. Add the high stakes, both climatic and financial, and there's certain to be real drama.

Still, it's a good idea, and if we're lucky, we're going to see an intense and detailed debate between proponents of global trading and proponents of development funds. And, in truth, the two wouldn't seem all that different to the people of the North: Under either system, countries would get their (basically per capita) allocations, overemitting countries would have to pay for additional environmental space, and the money would reappear later (we hope) as clean energy and development in the South.

In either case the rich must pay, and the money must be spent on clean energy and sustainable development. And one point must be stressed: *The alternatives to trading do not obviously solve the political and economic problems associated with emissions markets, and they have additional problems of their own.* No one, as far as we can tell, holds the moral or strategic high ground here.

We need to consider lots of options, and what really matters, in the end, is that the system be adequate, fair, and transparent. *Because the real problem is going to be convincing the people of the North that the money they have to pay isn't "aid," or "charity," or a "tax," but a payment for the use of resources that properly belong to the people of the South.* The real problem is going to be making the polluters pay, and the danger of the "trading debate" is that it threatens to obscure this central fact.

Keep in mind that it's likely to be a great deal of money. Perhaps not immediately, because right now rich-world subsidies are so out of whack—with billions upon billions of dollars being pumped each year into shoring up the fossil-fuel economy and obscuring the easy money available from efficiency—that the first years of a clean energy transition, once begun, may well be ridiculously cheap and easy. But the first years would pass, and later, when the emissions cap comes down, the price of emitting carbon will rise. Still, remember this: The rich will always have an easy alternative to buying environmental space from the poor. All they'd have to do is decarbonize their economies.

Seems fair enough to us.

A CRUCIAL PROVISO

Markets are devices, devices that today have been largely captured by the rich and the powerful, and our intention is hardly to hurry that process along. We agree, rather, with the International Forum on Globalization, which in a recent position paper, *A Better World Is Possible: Alternatives to Economic Globalization*,[5] has this to say about what it calls "common heritage resources":

> There is an appropriate place for private ownership and markets to play in the management, allocation, and delivery of certain common heritage resources, as for example land, within a framework of effective democratically accountable public regulation that guarantees fair pricing, equitable access, quality, and public stewardship.

This is well said and bears close reading. For even with the temperature rising and the world teetering on the edge of war, the "free traders" are pushing for a new round of global

Tom Athanasiou and Paul Baer

deregulation, and the corporations, relentless in their vigor, are pressing for the privatization of the water, the genome, and, of course, the air.

In this context, we have to keep our eyes on the ball. One way or another, a massive transition is coming. One way or another the air is going to become property, and the way forward, it seems to us, is by fighting to make it *common property*, property that we all share, equally, by virtue of our common rights as human beings.

The issue is not trading. The issue is forcing the rich to pay their share and preventing the privatization of the commons by establishing the institutions and politics of global public ownership. The issue is finding a just path to a just and sustainable world.

For no unjust path will get us there.

We cannot allow the warming to continue, but neither should we imagine that stopping it will be painless. There will be winners and losers, and some of us will lose a good deal more than others. Coal miners, whose jobs must disappear, are just the tip of the iceberg. One way or another, the price of carbon-based energy must greatly rise, and this means that—absent a comprehensive and visionary just transition program—many poor people will find themselves, once again, with the short end of the stick.

There's no end of ideas in the just transition[6] portfolio—community transition assistance, clean development, the

Sky Trust,[7] labor-friendly climate policies,[8] revenue recycling, domestic tradable quotas[9]—and it's long past time to put them on the table. Indeed, in a period when the just transition discussion should be heating up, it's wrapped instead in silence, and the fact that few people can really imagine governments funding just transition programs in anything like an adequate degree is clearly one of the reasons.

Which is, actually, where the loop closes. For if the atmosphere becomes common property, we'll be able to charge polluters for dumping carbon into it, and this money will, by most accounts, be sufficient to fund both clean development and adequate just transition policies. If we go the Sky Trust route, it will even fund annual payments to the citizen-holders of the atmospheric commons. But if the atmosphere is grandfathered or otherwise privatized, then, frankly, we'll be out of luck. Whatever else we accomplish, it will be too little.

The challenge, now, is to work our way forward in a way that's simultaneously pragmatic and visionary. It won't be easy, but just because the larger crisis is so vast and manifold, just because it encompasses not only global warming but also the ecological crisis in general, it should be possible. Along the way, we're going to have disagreements about globalization, markets, sovereignty, development, and much else besides. So let us, please, be clear: such disagreements, honestly engaged, can only be to the good.

The last thing we need right now is certainty.

Tom Athanasiou and Paul Baer

CHAPTER 7

THE FUTURE OF THE CLIMATE-PROTECTION COALITION

The Kyoto Protocol isn't just a climate treaty. It's also an economic treaty, and indeed it's the very first economic treaty that can plausibly be counted as a real step toward "sustainable development." It's weak, but it marks a path, one that we may yet be able to hold to.

Everything depends on how things go from here. Will Kyoto's trading and other flexibility mechanisms function well, or at least well enough? Will the Bonn coalition that in 2001 saved Kyoto from death by American repudiation manage to hold together? And, most of all, will Kyoto—as its supporters so fervently hope—open the door to long-term solutions?

When considering this last question, keep in mind that the long-term begins very soon indeed, with the debate over Kyoto's "second commitment period" scheduled to start no later than 2005. Yet even as eyes turn to the future, Washington, in its withdrawal from the Protocol, remains a dark and inescapable presence in the politics of the short-term. And this is unlikely to change anytime soon. When Republican strategists such as Stephen Moore, the president of a political action committee named the Club for Growth, can unabashedly assert, "I'm adamantly opposed to energy conservation. We're not running out. All we have to do is go out and find it and produce it," we know that the battle has taken on deep religious overtones, and not just on the environmentalist side.[1]

We should probably go on, at some length, about the Republicans. But, perhaps unwisely, we're going to assume that you already know the situation. The current U.S. administration clear embodies the smug and self-interested certainty of a right-wing cabal far too powerful to be taken lightly. When, ten years ago, George Bush-the-Elder announced that "Our lifestyles are not up for negotiation," he captured the same aggressive, know-nothing tone that his son used when he declared that the United States would not ratify Kyoto because it's "not in the United States' economic best interest."

And what else is there to say? We could talk about U.S. unilateralism, or the mania of consumption that so often seems to have taken over our souls. We could talk about the bombs, or the SUVs, or the corruption and political bribery that have become the standard operating procedure of the American corporate state. We could talk of strategic missile defense, and the false promises of militarily-defined security. But these are all just background, and we're going to assume that you already have it.

Tom Athanasiou and Paul Baer

Now, a thought experiment: Assume that we're right, and that there's no path out of here that doesn't begin by offering the South a fair and realistic development path. And assume, on the climate side, that this means a phased convergence to per capita greenhouse gas emissions entitlements. What then would this tell us about the politics of the way forward, and what, specifically, would it tell us about the role of climate politics in breaking the grip of the anti-environmental right?

A tough question, this one, but here goes...

First, consider the view from Washington. Many U.S. climate activists, particularly those based in its capital, see the overwhelming priority, just now, as "keeping a door open" for the United States (under a future administration, of course) to reenter the Kyoto regime. They speak for pragmatism, point to the weight of U.S. power, and see no possibility whatsoever that the United States will sign a climate treaty that doesn't satisfy the Byrd-Hagel demand to include the developing countries. They argue that at least some large developing countries must commit to limit their emissions, and that they must do so without raising the "rights issue" or otherwise throwing red meat to the Republicans.

We, for our part, want a regime that's both adequate and fair. We think this means engaging the issues of emissions rights and ecological debt, and doing so quickly, in the political opening that will follow Kyoto's entry into force, while we still have time to make it down some sort of soft-landing corridor. Still, we willingly admit that in the end, our personal desires are immaterial. The real issue, even ethically, is what will work, and in this we're more than sympathetic to the realist culture of the Washington environmental establishment, with these differences: We see the rights issue as paramount, and we don't see the United States as the lynchpin of the future.

What happens in Washington will of course be crucial. But the strategic problem, as we see it, is isolating Washington, or at least its current residents, rather than rushing to sacrifice a rare moment to their mad vanities. For when we look at the emissions projections, we conclude that the path taken by the South will be at least as important as the path taken by Washington. The question, as we see it, is how to construct a climate regime that will *really* interest the developing world, and, to that end, how to strengthen the climate-protection coalition that emerged in Bonn between the Europeans and the developing world, the coalition that saved the Kyoto Protocol.

And while we don't know the answer, we do have some thoughts.

THE LOGIC OF THE CLIMATE-PROTECTION COALITION

Recall, if you will, chapter 2, in which we argued that even a concentration target of 450 ppm CO_2 is unacceptably dan-

Tom Athanasiou and Paul Baer

gerous, and the subsequent discussion in which we compared the 450-ppm path to the IPCC's A1 "balanced" scenario, its de facto business-as-usual scenario. Our conclusion was that whatever happens, it's going to have to have *real force* behind it if it's going to bend the A1B curves down to the 450 path, let alone to the safer regions below. So if we're in fact on the A1B path, or anything like it, then we really don't have much time at all—which is why the next two decades will be so decisive, and why any strategy that doesn't recognize this will fail.

If we think big, however, we have a chance. For despite all the contradictory forces pulling upon the leaders and people of the South, many of them would gladly pursue genuine sustainable development if only they could. Unfortunately, this is quite impossible while the North, with the United States in the lead, continues to drag them down the path of development as usual. The details, at this point, are well known: structural adjustment programs that make socially and ecologically responsible development impossible, crushing and illegitimate debt burdens, World Bank project portfolios still dominated by coal and oil and gas, pathetically small amounts of "aid" (with lots of strings attached), and, of course, the World Trade Organization (WTO) and all the other institutions of "free trade," relentlessly deregulating the global economy, even as they drive Southern countries into an export-led development game that few of them can even hope to win.

This is the context within which the North-South climate showdown is going to finally take place, and only by understanding this can we see what climate politics must become. It comes to this: Global warming highlights the terrifying future now visible down the path of business as usual, and then goes one crucial step further to announce the peculiar, desperate power of the poor. Global warming forces us to see that even in a world dominated by the North and its corporations, *the South commands the vast power of historical inertia.* Its trump, quite simply, is that there must be massive cascades of clean-energy development in the South, because without them there will be a global ecological disaster. And that there must be large-scale institutional change, or these cascades will never arrive.

Here, then, is the challenge: As the severity of the expected impacts and the reality of the climate's sensitivity become better known, incrementalism alone is becoming manifestly implausible as a means of achieving a soft-landing corridor. We must, of course, find a path to early action, but not just any path will do. To claim seriousness, we'll have to explain exactly how we intend to make the massive emissions reductions that we now know are necessary. And perhaps paradoxically—for the roots of our predicament are unambiguously in the North—this means that the compass must now turn south.

Tom Athanasiou and Paul Baer

The South's grudging acceptance will not do. In fact, *there's no future for the climate negotiations unless they come to excite at least some large developing countries.* The future lies in a coalition between the European Union (EU) and the developing world, and many in Europe want desperately to secure it, but given the United States' power (and Europe's resurgent right), the EU will continue to vacillate. Thus, in marked contrast to the South's typical ambivalence about environmental treaties (which its politicians often dismiss as rich-world luxuries) a robust climate coalition will require the support and leadership of the South. And these will not flow from a strategy of denial.

Realism must follow the soft-landing path, which means keeping us under a low—and declining—global emissions cap. And rather than accepting the North's altogether inadequate "willingness to pay," our goal must be to increase that willingness by clarifying and demonstrating the North's "obligation to pay," its historical responsibility, its ecological debt. The point must be to make this debt into a major global issue, link it to the imperatives of sustainable development and climatic stabilization, and use it to pry loose the money needed for adaptation and compensation. The point, finally, must be to make it too politically expensive for American politicians and their allies to continue to refuse the discipline of an adequate climate regime and, thus, to break the deadlock.

What all this means, in practice, is that those of us in the North who want to make it down to a soft-landing corridor are going to have to do whatever it takes to make that corridor attractive, really attractive, to the South. The logic here must be crystal clear, so we're going to spell it out:

1. We must insist upon the North's historical responsibility for climate change. That is, we must show that the North, by overusing the atmospheric space, has made it necessary to move extremely quickly if the South is to develop without setting off a catastrophe. This is crucial, because if "the North is responsible," it's no less responsible if the costs—including the costs of adaptation—turn out to be high. Just the opposite, in fact. The greater the costs, the greater the North's responsibility for them.

2. Achieving a soft-landing corridor is not going to be cheap, because it has to happen fast. Moreover, the total cost of making it to a soft-landing corridor is not primarily a function of environmental rights but of the distance that must be quickly crossed. To be sure, there are plenty of "no regrets" and low-cost alternatives, but costs will nevertheless be high in some sectors, and for some nations. In fact, the need for rapid leapfrogging in the South will probably ensure that aggregate costs are high. And because costs will rise with the ecological adequacy of the climate regime, the "cost issue" simply can't be ducked.

3. Energy efficiency will not suffice. Inefficiencies, to be

Tom Athanasiou and Paul Baer

sure, must be eliminated wherever and as quickly as possible, but the critically "underdeveloped" parts of the world must "leapfrog" directly to low- and zero-carbon technologies. In fact, New Delhi's Centre for Science and Environment has convincingly argued that, particularly in the South, too strong a focus on efficiency could have the effect of *locking out* the renewable revolution.[2]

4. Despite the need to stress the North's responsibility, we argue for a rights-based rather than responsibility-based politics of climate equity. So note that from the point of view of ecological debt, a rights-based transition would let the North off easy. A responsibility-based deal would lead to even higher payments from the North to the South, but we just don't think it's going to happen. A rights-based deal, on the other hand, may well be possible, and soon, particularly if it's understood as a critically necessary compromise.

5. A joint, North-South campaign for such a compromise would have a number of extremely positive effects. More on this in the next chapter, but here's the headline: The South will only embrace the sustainability agenda if it promises a *real development strategy.* And with Kyoto's entry into force, as we move into the debate about impacts and equity and developing country commitments, we're going to have a rare opportunity to put just such a campaign together.

THE BOTTOM LINE

Sustainable development—the real thing and not the rhetoric—will come at a price. Agenda 21, the "comprehensive plan of action" that was the principal product of 1992's Earth Summit,[3] estimated total additional financing needs for sustainable development at over $600 billion a year.[4] This is real money and, of course, nothing even remotely like it has been forthcoming. Indeed, even "aid" budgets, such as they are, continue to drop.

We may, however, have finally won a major battle. For despite all its faults, the Kyoto Protocol, reduced by chains of compromises into "Kyoto Lite," may yet suffice as a first step, and thus allow another. Further, the optimistic illusions of the early post–Cold War period are now stale, even bitter, history. We may now, if we wish, see our real conditions of life, and given this, a great deal is possible. Given a visionary climate treaty, we may yet see the North-South coalition we need to make sustainable development real.

And what of the risk that too much "rights talk" will enflame the right wing, and make U.S. reentry into Kyoto impossible? We see it as a risk worth taking. The problem of the United States, after all, isn't only a problem of the Republicans. It's also the problem of a proud people who find themselves, with every passing year, farther from their fading ideals, the problem of a people with difficult choices to

Tom Athanasiou and Paul Baer

make. In this context, we should not begrudge them the truth. Our message should be something like, "If we're going to take more than our share, we should at least pay for it," and it's a message, you'll note, that requires that we be able to define "our share."

One last point: There's a pattern in global negotiations, one in which matters of life and death are watered down to the point of absurdity, just to keep the United States "engaged." It's old news in the climate talks, but it was just as clearly visible at the International Conference on Finance for Development held early in 2002 in Monterrey, Mexico. Chalk Monterrey up, then, as another lost opportunity, for the price of U.S. participation was heavy indeed. All talk of meaningful changes in the development financing system—eliminating some of the trade barriers that discriminate against Southern exports, or introducing an international development tax—were pulled from the final agreement. And to what end?

So that George W. Bush could attend.

A lesson must be drawn. Eventual U.S. action—or inaction, as in the case of global warming—is generally not worth the price. Instead, it's time to start talking, in very concrete terms, about ways forward that are both equitable and adequate. And just now it looks like that conversation will be a lot easier with the United States on the sidelines.

CHAPTER 8

GLOBALIZATION AND GLOBAL WARMING

The summer of 2001 might almost be ancient history, belonging as it does to the world before September 11. But recall, if you will, an instructive moment from that long-past July, as it unfolded during two simultaneous intergovernmental conferences. The first, the G8 summit in Genoa, saw a sudden flash of stark reality as, amidst massive swirling protests, an Italian cop shot and killed an over-exuberant "black bloc" protester. At the second—the emergency Bonn climate summit that followed 2000's deadlock conference at The Hague—there were also police and protesters, but just about everything else was different. The police were well-trained Germans, and the protesters, hostile though they were to the assembled "bureaucrats," were nonetheless acutely aware of the importance of the negotiations. And, indeed, quite in contrast to the mock statesmanship of the G8, the Bonn meeting saw the ministers of 178 nations locked in tense all-night negotiations aimed at saving a besieged Kyoto Protocol.

The two events were clearly tied together, most manifestly by the calls and communiqués that were shuttling between key ministers in Bonn and their chiefs in Genoa. But there was more as well, much more. Taken together, they portrayed the Janus face of globalization: on the one side, the domination of international governance by the core capitalist countries, their corporations, and their central banks, and with it the maintenance of a firm and sometimes brutal wall against protest and creeping delegitimation; on the other, a rare victory, and the opening of a new chapter in the story of multilateral governance, one featuring increasing participation by both the developing countries and the international NGOs, one in which the protest and policy cultures merge, one that suggests a way forward.

Two flash points, then, and a rapidly changing landscape. For even at Genoa, long before September 11, the "anti-globalization movement" was morphing into a new movement that knows that "globalization" is far too vague a target, that the real battle is and must be for justice. And so, too, on the other side, where the climate movement, faced with the need to go beyond Kyoto, is being compelled to its own rethinking, and its own confrontation with equity. The two movements, finally, are being thrown more closely together; just as clearly, the result will be something new.

A NEW KIND OF MOVEMENT

There have always been huge overlaps between the climate and globalization (formerly anti-globalization) movements.

Both strain to rework the terms of global finance. Both struggle against recalcitrant and self-interested elites. Both focus on the same essential problem of sustainable well-being on an ecologically overburdened planet. Both have long been global-justice movements, in deed if not in name, and both, today, confront the challenge of justice in new and more explicit ways.

The old environmentalism is dead; Malthusianism is slouching away toward the dustbin; facts are being faced, conclusions being drawn. The rich consume massively more resources and produce massively more pollution than the poor, and this, quite inescapably, is the baseline of both true radicalism and honest realism. More precisely: We cannot hope to find justice in a world where the poor come to live as the rich do today, for there is not world enough. There will have to be another kind of solution. There will, indeed, have to be new dreams on all sides, and the rich, in particular, will have to make those dreams possible by learning to share.

The emergence of a self-conscious global-justice movement is the best news in years, but there's lots of work to do. For one thing, most all the street-side effort has gone into confronting the free traders and the development banks. It's been great, but we need more focus on international finance, per se, and a more specific, and strategic, discussion of development and sustainability. This is coming, though, and just

Tom Athanasiou and Paul Baer

in time. And as it does, increasing attention will turn to the hothouse of climate politics.

It may come as a surprise to the globalization folks—who earned their spurs in long and spirited campaigns against the World Bank and stood, in Seattle and elsewhere, for democracy, labor rights, environmental protection, and the broad empowerment of those at the bottom—that they need the climate people, with all their elite scientific, institutional, and economic preoccupations. And yet they do.

The Kyoto Protocol was a watershed in the battle for an ecological economy. It's not merely that with Kyoto's entry into force, carbon, or rather the right to emit carbon, will finally have a price. It's also that this price will have been imposed by *an open multilateral process based in the United Nations*, a process that was held together despite the best efforts of both the Bush administration and the fossil-fuel cartel to destroy it. With the world desperately in need of democratic global institutions and the government of the United States, if not its people, lost in dreams of imperial unilateralism, the significance of such a step should not be underestimated.

The climate movement was born from the intensely research-based culture of atmospheric science, and it exemplifies both the strengths and weaknesses of science-based environmentalism. Consider the SRES scenarios, which we've cited many times. With all their vividly specific quantitative perspectives on both transition paths and ecological limits, they're hardly the works of the romantic misanthropes who populate the mists of anti-environmental mythology. And, frankly, the globalization folks could benefit from taking a close look at them.

Yet the climate people suffer their own naiveté, particularly when it comes to the coming collision between the world of multilateral environmental governance and the institutions and ideology of "free trade." Ask the average climate policy wonk if the greenhouse negotiations are on a collision course with the World Trade Organization and they'll probably say no, that if national environmental laws are written carefully there should be no problem. But this, as it happens, is a strange and almost willful illusion. To be sure, there may be a parallel world in which everyone cooperates to stabilize the ecosystem, and in it, no doubt, everything fits nicely together. But in this less fortunate world, in which the big transnationals are more powerful than most governments and the oil companies more powerful than most transnationals, in which globalization has radically reduced the abilities of national governments to regulate cross-border capital transfers, in which the United States and not some "unimportant" edge-case country like Albania or Myanmar is the defining "free rider," things are just a bit different. In this world, it will be difficult indeed to avoid a trade-environment train wreck.

Tom Athanasiou and Paul Baer

The new wave of economic globalization has already made it harder to fight poverty, raise social standards, or protect the environment. Companies threatened with social or environmental legislation can and do move their investments elsewhere, and governments everywhere know it. Some turn timorous before the threat of capital flight and some just become aggressively "pro business," but either way, even powerful countries hesitate to pass or enforce strong social and environmental rules. And there are no strong international regulations to take up the challenge.

And that's today. Tomorrow, things may be much worse, particularly if the free traders succeed in moving their "new themes"[1] through the WTO agenda. These include the opening of services, and particularly energy services, to international competition, the prohibition of national restrictions on international investment, and the curtailing of the ability of national governments to protect national monopolies. It's an aggressive agenda, and it would be quite enough, when added to an already oppressive trade regime, to prevent most governments from enacting carbon taxes, domestic cap and trade systems, or other potentially effective domestic carbon-control schemes.

Unless, that is, countries and groups of countries were able to insulate themselves from the global economy and all its many free riders. This they could do so easily enough if—and it's a big if—they were free to complement domestic carbon emissions regulation with tariffs or border taxes designed to discourage competition by cheaper imports not subject to similar controls. Alas, such mechanisms already, according to one analyst, "raise complex questions with respect to WTO consistency and the conditions under which border taxes can be adjusted to accommodate a loss of international competitiveness."[2]

If the free traders succeed in their next push, the answers to these "complex questions" will become altogether too obvious, for climate-related border taxes will be defined as unequivocally "trade illegal." And this, to be very clear, must not be allowed to happen. Even the "complex questions" that today trouble the interface of trade and environmental law must be cleared up, and to one very simple end: *Multilateral environmental agreements, as well as human- and labor-rights law, must unambiguously trump the WTO and its trade rules.*

MORE POWER TO THE SOUTH!

Our point, in case you haven't guessed it, is that the linkages between climate and other global issues—trade and development first among them—promise to raise the climate-protection coalition, and the developing countries within that coalition, into a remarkably strong position. Look at it this way: The people of the South are becoming restive, and everyone knows it. The only way forward, moreover, is that

Tom Athanasiou and Paul Baer

strange process we know as "development," and we had all better do our best to make sure that the sustainable variety is both real and available.

Realism, however, compels us to admit that "sustainability" is not the direction of history. Think of 1992's Earth Summit, and of how few of its promises have been redeemed, and the case is clear: The South will not easily take a low-carbon path, for it is not open. Leave aside the crashing wave of neoliberal integration that followed the collapse of the Soviet bloc, and the trade and aid politics of the North, both of which actively encourage carbon-intensive growth. Forget, too, that the fossil cartel controls much of the world. Consider only that "creative destruction" is perhaps the most remarkable and enduring feature of capitalist development, and that the South, weak within global tides, is most easily destroyed. You will see why Southern countries cannot be expected to pursue dreams.

What it can be expected to do is to pursue its own interests. Thus, as a bloc, the South will fragment or not, based almost entirely on those interests. As a set of individual states, each competing with the other, it will step toward a more efficient form of modernization, or not, based almost entirely on the logic of short-term politics and profitability. To imagine otherwise is to hope for statesmanship and vision, and perhaps these will come. But even so, Southern diplomats will not make history under conditions of their own choosing. Rather, economic and ecological crises will continue to evolve, and the climate crisis will in many ways set their terms. With Kyoto in effect, the debate will turn, as it must, to the preconditions of a regime that can bring China, and India, and even the United States into the fold. In all this, the South will play its cards as best it can.

Will this lead to deadlock, or even disaster? Perhaps. The South has its full complement of madmen, fools, and dictators. But so too does the North, and in any case, we've no choice but to seize this opportunity. For while the South must develop along a low-carbon trajectory, today's global institutions, all of them controlled by the North, continue to force it into paths of fossil-fuel dependency and desperate, export-led industrialization. This will clearly have to change, and until it does, Southern elites will find it remarkably easy to treat the North's interest in environmentalism as a scheme designed to suppress their development. Hypocrisy breeds more hypocrisy, and Southern development, after all, *is* being suppressed. Northern environmentalists make a most convenient foil, particularly while they remain complicit.

Could a visionary climate plan, anchored in an alliance between the EU and the South, shift the field? We believe that it could. We believe that if the environmental movement, North and South, stood together for *a just climate regime designed to ensure Southern development* (the real thing this time), it would become far more difficult for bad actors, Northern or Southern, to play their usual game of divide and conquer. We believe that a climate deal that *actually promised to fund sustainable development* would predictably split the Southern elites, and in exactly the way we want to split the elites everywhere—between those interested in the health and development of their peoples and those who care, in the end, only for their own vanity. We believe, moreover, that if climate protection became a drive for genuine sustainable development, then even self-interested Southern elites would develop a keen interest in it, and for the most practical of reasons: The alliances it would engender would offer them real and tangible benefits in their areas of greatest concern, areas like trade and investment.

Why, when the worst impacts of climate change will be in the South, do its leaders continue to equivocate? No doubt there are many reasons, their privileged positions chief among them. But try, for a moment, to see the world through Southern eyes. See that it is one in which the Northern powers don't even keep their promises of "aid," and that while the entire Northern political class poses as "free traders," the estimated cost of Northern trade barriers to Southern economies amount to well over U.S. $100 billion a year, far more than the North grants the South in development assistance.[3] See that even the Europeans, though often allies of the South in the climate talks, are only business-as-usual players when their core economic interests are at stake. See that, time and again, promises of development have turned rancid in the act.

It comes to this: The South will only embrace the sustainability agenda when it promises a *real development strategy*, one that Southern leaders genuinely believe can find traction even in a world like ours. Until that day, any really adequate climate treaty will remain out of reach. Until that day, the South, like the United States, will refuse any international environmental regime with strong and enforceable sanctions. Until that day, Northern environmentalists will remain, as today, frustrated by a developing world curiously uncommitted to large-scale climate cooperation.

Which brings us back to our case for a rights-based treaty. For honest sustainable development will demand money, and regulations, and trust, and none of these will come without a bold stroke. Nor, to repeat a key point, will adequate early action against global warming ever be possible, not unless the North pays for it, and on a scale that can only be defended in terms of our common humanity, and our common rights to the Earth's common spaces.

Tom Athanasiou and Paul Baer

CHAPTER 9

THREE FUTURES

It's quite a challenge to be simultaneously pragmatic and visionary, not least because when pragmatism takes the stage, the "visionary" wins polite denigration and quickly fades into the background. This time, however, the vision—of a *rights-based* climate-protection regime—must not be allowed to fade, and for one emphatically pragmatic reason above all others: In its absence, pragmatism will become politics as usual—accommodation to power—and the atmosphere will become merely the property of the rich. And this, to connect the dots one last time, would mean the loss of all legitimate hope. For if the atmosphere becomes the property of the rich, it will be the end of far more than the climate negotiations. It will be the end, as well, of any chance for a soft landing in which we avoid large-scale atmospheric destabilization, greatly accelerated extinction rates, and an extremely high-stakes passage into an age of global resource wars. In such a context, a climate deal, if there was one, could not possibly hold.

The details here would fill whole shelves of books, but that, really, is not the point. Better to make a leap, to look forward to a future in which the politics of trade and of trading, of climate and of justice, of peace and of democratization all somehow make sense together. Better to stipulate in advance that *a new world is possible*,[1] that its demands and contours are becoming clearer, and that we don't have forever to bring it into existence. Better, in fact, to do all this quickly, and then to turn back to sustainability, and to climate change in particular, and ask what must be immediately done.

Some things are obvious. We must, for example—and this is very much in the old "anti-globalization" mainline—force the World Bank and the other international financial institutions to stop financing fossil-energy development.[2] Beyond this, the power of the fossil-fuel cartel must be broken, and as soon as possible; this is clear, and of course far easier to say than to do. And beyond this, in turn, lies the global-justice movement's whole expansive agenda, large and intimidating in its difficulty. There's debt reform, and the need to end neoliberal structural adjustment. There's the threat of emerging diseases, and the need for "basic rights" to clean water and sanitation and food. There's the need for a new trade regime, and a new realism about war and peace, poverty and security. All of this will remain, undiminished or even increasing in importance, as we continue our long, slow reckoning with the demands of climate protection.

Tom Athanasiou and Paul Baer

But climate change adds a twist. It brings things together and highlights the bitter, desperate power of the poor, the difficulty of escaping the channels of our historical and institutional "path," the costs of failure. All of which makes it easy to conclude that climate change may turn out to be strategically decisive, which is just why, particularly in a book like this, it's crucial to stress that climate isn't the whole story. We propose to do this with our own version of the Global Scenario Group's "three futures" analysis.[3] The idea here, in a nutshell, is that we're at a world-historical turning point, and that, broadly speaking, the world can go in one of three different directions. Like so:

THE BUSINESS AS USUAL FUTURE

The BAU world is today's world, but reformed enough to hold together.

Market-driven development reigns, and "free trade" and globalization continue more or less as usual, but the worst excesses of the neoliberals are constrained by policy reformers and by elite fear of the poor. There's an ever-greater awareness of the social, technical, and political alternatives, but despite incremental moves toward fairness and sustainability, they remain marginalized. A few green technologies, like wind power and hybrid cars, are extensively developed, but the larger patterns of social and technological development are constrained by the institutional conservativism of the rich and the powerful, and by our collective inability to solve the problem of class.

The transnationals become increasingly dominant. The climate rapidly changes, but the rich-poor divide prevents any effective response, so the rich and the poor both "adapt," each in their own way. Social brutality increases, but it is accepted as inevitable. Ecological decay continues apace, and significant local and regional ecosystems collapse. "Individualist" consumer culture continues to spread, and cultural diversity continues to erode under the pressure of a corporate monoculture. Traditionalist and fundamentalist backlash is contained, but the world becomes a grimmer, less friendly place, haunted by a sense of irretrievable loss.

Pessimism becomes the norm, but somehow, still, the "center" continues to hold. There is no uncontained social explosion. Always, there are other, better, possibilities, but deadlock rules the day. Farsighted factions of the elite class issue ever-grimmer warnings, but still, reforms are limited to small ameliorations and insecure safety nets. It can't go on, but it does. For a while.

BARBARISM AND THE FORTRESS WORLD

Here, the BAU pattern collapses into an even more Hobbesian one.

In the face of accelerating social and ecological decay, it becomes clear that nothing effective is going to be done. The

Tom Athanasiou and Paul Baer

planet warms, and the impacts rise to horrific levels, but with the elites unwilling to face the demands of redistributional equity, the climate-protection coalition fragments and collapses. New and terrible kinds of wars take center stage, dominating international politics and solidifying the deadlock.

In hindsight, it becomes clear that an opportunity has passed. Hope fades. Environmental and social deterioration begin to feed on each other, pulling the world into a self-reinforcing downward spiral. Governments, largely captured by private interests, withdraw from social projects, as economic and racial divides are accepted as normal. Economic polarization continues to increase; the sense of loss deepens; culture becomes more commercialized, more cynical, and more decadent than could have previously been believed; pessimism fades into barbarism. The global poor become restive, desperate, dissatisfied by images of affluence, and angry cabals emerge to strike out against the rich. Messianic violence increases on all sides, but it only reinforces the fortress pattern. Everywhere, the rich withdraw into strongholds. The richer regions of the world erect ever-higher barriers to immigration. New schools of "realism" emerge to repeat, in a thousand different ways, that nothing can be done. Xenophobia and garrison culture become the norm. Food insecurity and disease dominate the lives of billions of people. Poor countries fragment and deteriorate. The global economy sputters, international institutions weaken, and the affluent see that they will eventually be engulfed. Still, the party of force silences the few who believe that reform is possible. For a while, eco-fascism, an authoritarian form of sustainable development, becomes the hope of the affluent. But it doesn't really work, and everyone knows that it won't last.

THE GREAT TRANSFORMATION

Alternatively, we could seize the opportunities while they're still open.

In fact, both the shock of 9/11 and global fear of a neo-imperial United States could help to catalyze the change. In the end, though, a great transformation demands a new global politics, one that melds the imperatives of poverty alleviation and ecological protection together with traditional and radical democracy to yield something strong and fine and new. The Bonn coalition, in which Europe and the developing world joined together to save the Kyoto Protocol, could turn out to prefigure such a new international politics–if it survives.

It's been obvious for years that some sort of wake-up call was needed, and it does seem that the new millennium is doing its best to provide one. Much will depend, now, on how the conclusions are drawn: whether a significant fraction of the elite classes come to support global justice, or if they all stay mired in the BAU world; whether the UN system finally comes into its own, or remains forever subsidiary to the ethos and institutions of international business; whether we actually set out to protect the climate, or only pretend to do so.

We know, in any case, that we're on a crucial cusp. The green energy revolution could come in earnest and catalyze a larger technology revolution that, together with sustainability culture and left/green economics, brings a new kind of low-impact prosperity into being. The global market could be reframed by a new global democracy and constrained within social and environmental limits. We could face, finally, the problems of race and class. Democratic culture could return from the brink, and consumerism could lose its grip on our imaginations. Equity and efficiency, together, could define a new ethos. "Sustainable development" could become more than tired, disappointing rhetoric.

It's a long shot, perhaps, but it's a whole lot better than the alternatives. And, frankly, it's bracing to stare those alternatives in the eye and draw conclusions. In fact, after 9/11, drawing conclusions, connecting dots, and perhaps even "understanding" itself are risky political acts, particularly if your conclusions are subversive ones like, say, "as you sow, so shall you reap."

Hope is, as they say, a duty, but it isn't an easy one, not these days. Whatever it is that we the rich have been sowing, the crop will come in time. It always has in the past, and there's no reason to believe, this time, just because the stakes are so very high, that things will be different.

CHAPTER 10

A FEW LAST WORDS

The "atmospheric commons." Even to speak the phrase is to invite the classic reply, pioneered in Garrett Hardin's *The Tragedy of the Commons*, that common property is doomed to overexploitation and destruction. But this, we're happy to report, is confused nonsense. In fact, of all the problems with Hardin's famous screed, the worst is its title. A "commons" is a socially regulated space, in which both rights and responsibilities are clear, whereas Hardin was describing "open access" regimes in which anything goes, especially for the rich and the powerful. He should have called it *The Tragedy of Unregulated Open-Access Resources*.

Our argument, at this same level of abstraction, is that if we honestly intend to protect the atmosphere, then we have to fight to convert it from an open-access resource into a commons, a limited, socially regulated global commons in which access is apportioned to us each in equal measure, by virtue of nothing more or less than our common humanity.

Easier said, of course, than done. But nevertheless it must be done, for contrary to the assurances of the economists, the market isn't going to solve the problem of scarcity, or make it unnecessary to create a global commons. The market, rather, will have to be democratically regulated, and for real this time.

The economists tell us that when a resource becomes physically scarce, its price goes up, and scarcity, thus transformed into opportunity, becomes a problem to be solved by entrepreneurs, technologists, and just plain budget-minded people. Substitutes are found, inventions are made, demand goes down, and equilibrium returns. Scarcity, in this classic cornucopian story line, is the mother of invention.

And, in fact, the story line is reasonably true—for some resources, some of the time. But it's not true in general, and it's particularly not true when scarcity comes to open-access resources such as the atmosphere, or indeed the oceans, the freshwaters, and what little remains of the frontier. In such cases, where resources are "just there for the taking," economics as we know it quite fails to play its much-advertised role. This is a long story—and an important part of ecological economics—but suffice it to say that, just now, there are two ways forward, and that both of them involve that peculiar social institution known as property.

The first is that open-access resources, now scarce—think of the rivers, or prime real estate, or, in fact, the air—can be *privatized*. This is the way of the world, the modern world in any case, and it has generally sufficed to solve "the tragedy of the commons," though only at the cost of another tragedy, that of the haves and the have-nots.

The second, a bit more speculative, requires us, in effect, to reinvent the commons by inventing new practices and institutions of *common property*. This, too, is a long story, and one that's still being written. But we can say this: Common property is very different from private property, and its ultimate justification must come in terms of public rather than individual rights. Further, when it comes to the global commons, those public rights must ultimately take the form of equal access, and equal shares. Nothing else makes, or should make, any sense.

This may all sound pretty abstract, but it's actually very simple and concrete. A park is a commons, and there is nothing particularly mysterious about a park. It's just a bit of land, usually a nice one, that's been protected socially rather than privately and, indeed, protected from privatization. No one owns it, but it isn't open access either; parks come along with rules and customs, and these rules and customs are enforced. That's the whole point!

A commons is a socially regulated space that is "held in common." It's a fascinating institution, now finally beginning to get the attention it deserves, and, if we're correct, it's also the key to the solution of the climate crisis. The problem of the atmosphere, after all, is the problem of fair global regulation. And *Dead Heat*, in the end, is only an argument that such regulation will be impossible to win and then enforce unless it is grounded in and legitimated by a common property regime that includes use rights, economic rights, and, of course, decision-making rights.

Rights, then, are not abstract things, and if we think they are, this is perhaps a sign of our confusion. Rights are deadly real, though they tend to be more visible to those without them. Thus, you may own many things, but you may not own another human being. That's what it means to put limits on the "right" of private ownership, and that's the sort of thing that's at stake here.

* * *

Finally, as we don't wish to be remembered for making airy and idealistic claims, we will end by restating our core strategic argument:

We don't have much time, not if we want to make it to a soft-landing corridor, and there are good reasons to doubt that we'll make it under anything like the existing regime. Unless something changes, there's little prospect of "adequate early action." In fact, the future, in case you haven't noticed, is looking just a bit too much like "Fortress World."

We may yet change course, and looking into the abyss may be just the motivation we need. But we're not going to make it unless we set off, quite explicitly, to build a fair

Tom Athanasiou and Paul Baer

world. In this sense, global warming is as much an opportunity as it is a threat, for it presents us with what Herman Daly, the dean of ecological economics, once called "an optimal crisis." The global-warming crisis is big enough to get our attention and, perhaps, to focus our minds, but it's nevertheless small enough to deal with.

All we have to do is face the facts.

RESOURCES

ORGANIZATIONS

CENTRE FOR SCIENCE AND THE ENVIRONMENT (CSE)
www.cseindia.org/

India's CSE, well known among Southern environmental NGOs, has been enormously influential in the climate equity movement since 1990's publication of Anil Agarwal and Sunita Narain's *Global Warming in an Unequal World.* Their whole (extensive) site is well worth exploring, but their Climate Campaign, at www.cseindia.org/html/cmp/cmp33.htm, and their newsletter, *Equity Watch,* at www.equitywatch.org/, are of particular interest.

GLOBAL COMMONS INSTITUTE (GCI)
www.gci.org.uk/

The indefatigable Aubrey Meyer of London's GCI has been working for over a decade to put "contraction and convergence" onto the international agenda, and he has had some success, particularly in Europe. For GCI's long history with the issue, see www.gci.org.uk/papers/GCIArchive1989to2002.pdf.

CLIMATE ACTION NETWORK (CAN)
www.climatenetwork.org/

The Climate Action Network is a global coalition of over three hundred independent organizations working the climate issue, and many of the other organizations on this list are members of it. See, in particular, its *ECO* newsletter (www.climatenetwork.org/eco/), which CAN publishes from all major international climate meetings; read it carefully and you'll soon understand what we're up against.

The United States Climate Action Network (USCAN) Web site (www.climatenetwork.org/uscanweb/uscanhome.htm) provides a portal to the climate pages of most of the major U.S. environmental groups.

ECOEQUITY
www.ecoequity.org

EcoEquity aims to bring the international discussion home to the United States. We publish Climate Equity Observer (subscribe by e-mailing ceo@ecoequity.org) and we are, of course, the authors of this book.

REDEFINING PROGRESS (RP)
www.rprogress.org/programs/climatechange/

Redefining Progress has done outstanding analysis of climate equity from a U.S. domestic perspective, and has been a key facilitator of coalition building on the issue.

THIRD WORLD NETWORK (TWN)
www.twnside.org.sg/climate.htm

Long a major player in the anti-globalization movement, the TWN has added climate to its agenda and produced a variety of interesting analyses.

ENVIRONMENTAL JUSTICE AND CLIMATE CHANGE INITIATIVE (EJCC)
www.ejcc.org/

The EJCC Initiative is designed to bring U.S. environmental justice groups together to engage the climate issue. Check out the site and you'll see that the match makes excellent sense. EcoEquity is a member, by the way.

RISING TIDE
www.risingtide.nl/

Rising Tide is an international coalition of radical climate groups, strongly supportive of climate equity but skeptical of the Kyoto process. See also the British Rising Tide site at www.risingtide.org.uk/.

CORPWATCH
www.corpwatch.org/

The Corpwatch climate campaign is aligned with the traditional U.S. environmental justice movement, and this page is a useful window into its evolving take on global climate justice issues. The special focus here is on the fossil-fuel cartel.

THE U.S. SKY TRUST
www.usskytrust.org

The Sky Trust would give every U.S. citizen a per capita share of the revenue from auctioned emissions permits, and it would provide a serious Just Transition fund to assist workers and communities impacted by the inevitable economic restructuring.

THE TELLUS INSTITUTE
www.tellus.org/

Tellus is a small but influential think tank that has done outstanding work on energy and sustainable development and has a growing influence on the climate equity movement. Tellus is also a partner in the Global Scenarios Group; see www.gsg.org/.

CHOOSE CLIMATE
www.chooseclimate.org/

The brainchild of scientist/activist Ben Matthews, this site features a sophisticated, interactive climate model that can be used to explore emissions allocations scenarios, as well as a carbon calculator for estimating air-travel emissions. Check it before your next conference jaunt!

INTERGOVERNMENTAL AND NATIONAL BODIES

INTERGOVERNMENTAL PANEL ON CLIMATE CHANGE (IPCC)
www.ipcc/ch

The IPCC is *the* scientific authority on climate change. All of its major reports are available on its Web site; see notes throughout the text for specific references.

UNITED NATIONS FRAMEWORK CONVENTION ON CLIMATE CHANGE (UNFCCC)
www.unfccc.de/

The UNFCCC's Web portal hosts all the documents from the official climate negotiations, as well as a variety of useful related materials. And, at www.unfccc.int/resource/conv/index.html, it contains the full text of the treaty itself. The Framework Convention is already law, even in the United States. George W. Bush's father signed it in 1992, and it has been ratified by the U.S. Senate.

THE KYOTO PROTOCOL

For those of you who just have to check Article 17, the full text of the Kyoto Protocol is online at unfccc.int/resource/docs/convkp/kpeng.html.

Tom Athanasiou and Paul Baer

The EPA's climate Web site provides a good introduction to the issues, as well as scattered insights into the U.S. administration's rather conflicted position on climate change.

BOOKS AND ESSAYS

Climate Change Science: An Analysis of Some Key Questions

This notable 2001 report from the U.S. National Academy of Science really sticks it to the climate "skeptics." Available online at books.nap.edu/html/climatechange/.

Great Transition: The Promise and Lure of the Times Ahead

The original elaboration of the scenarios we discuss in chapter 9 (Boston Center of the Stockholm Environment Institute and the Global Scenarios Group, 2002). Downloadable at www.gsg.org/.

Green Politics and Poles Apart

The two volumes of CSE India's long and excellent overview of global environmental negotiations. *Green Politics* contains a fine history and overview of the climate negotiations, which was updated in *Poles Apart*.

To order, see www.cseindia.org/html/eyou/geg/publications_gen2.htm.

The Heat Is On

In many ways, this book by Ross Gelbspan (Addison-Wesley, 1997) is the classic treatment of climate politics in the United States, with lots of detail on the climate skeptics and their funding. See also Gelbspan's Web site, www.heatisonline.org/.

Who Owns the Sky?

A book-length introduction by Peter Barnes (Island Press, 2001) to the Sky Trust proposal for an equity-based U.S. climate policy. And see our review at www.ecoequity.org/ceo/ceo_3_3.htm.

"Equity and Greenhouse Gas Responsibility in Climate Policy"

This essay by P. Baer et al. (*Science* 289 (2000):2287) is a short, sweet summary of the arguments for equal per capita rights. Downloadable at www.ecoequity.org/docs/science.pdf.

"Seeking Fair Weather: Ethics and the International Debate on Climate Change"

A classic academic treatment by Michael J. Grubb (*International Affairs* 71 (1995):463–96) of the ethics of various proposals for emissions reductions.

"International Justice and Global Warming"

For those of you with access to an academic library, this is an outstanding review by Matthew Paterson of equity issues from a political philosophy perspective. In *The Ethical Dimensions of Global Change*, ed. B. Holden (New York: St. Martin's Press, 2001), 181–201.

"Subsistence Emissions and Luxury Emissions"

Henry Shue, a leading philosopher, examines a critical issue in climate justice. In *Law and Policy* 15 (1993):39–59.

"Equity, Greenhouse Gas Emissions and Global Common Resources"

A philosophical argument by Paul Baer for equal rights to global common resources. In *Climate Change Policy: A Survey*, eds. S. H. Schneider, A. Rosencranz, and J. Niles (Washington, D.C.: Island Press, 2002).

PORTALS

INTERNATIONAL INSTITUTE FOR SUSTAINABLE DEVELOPMENT (IISD)
www.iisd.org/.

Canada's IISD hosts both Climate-L, the premier list for announcement of climate-related activities and publications, and the *Earth Negotiations Bulletin*, a dry but objective review of the daily activities at major global environmental negotiations, as well as the Climate Change Knowledge Network, an international network for education and capacity building: www.cckn.net/

CLIMATE INDEPENDENT MEDIA CENTER
www.climate.indymedia.org/

Online independent coverage of climate negotiations, protests, and related events, as well as an archive of historical reporting.

GLOBAL CHANGE
www.globalchange.org/

Online review with links to a wide variety of climate-related publications.

CLIMATE ARK
www.climateark.org/

Extensive links to a wide range of publications on climate and energy policy.

PACIFIC INSTITUTE
www.pacinst.org/ccresource.html

A comprehensive and well-organized set of links to all aspects of the climate issue. It even has a subsection (3.3) on climate equity!

Tom Athanasiou and Paul Baer

NOTES

CHAPTER 1: AN INTRODUCTION

1. The Global Scenario Group's Web site is at www.gsg.org.
2. C. Wright Mills, *The Power Elite* (Oxford: Oxford University Press, 1956), 356.
3. The Kyoto Protocol will enter into force ninety days after it has been ratified by at least fifty-five countries, representing at least 55 percent of the 1990 Annex 1 (developed country) emissions. As we went to press, ratification by Russia would push us across the threshold.
4. The quantitative caps established in the Kyoto Protocol set limits on greenhouse gas (GHG) emissions from the developed countries only, and only for the years 2008–2012; this is the so-called first commitment period. The Protocol explicitly states that the negotiations for the second commitment period—whose dates have not been established but are widely assumed to be 2012–2022—must begin by 2005 and be completed by 2008.
5. See, for example, Kevin Baumert, Ruchi Bandhari, and Nancy Kete, "What Might a Developing Country Climate Commitment Look Like?" available at www.wri.org/climate/develop.html. And see *Building on the Kyoto Protocol: Options for Protecting the Climate*, edited by Kevin A. Baumert with Odile Blanchard, Silvia Llosa, and James F. Perkaus, (Washington DC: World Resources Institute, 2002).

CHAPTER 2: THE SCIENCE CHAPTER

1. The IPCC is a story unto itself; its own Web site is a good introduction, www.ipcc.ch.
2. *Climate Change Science: An Analysis of Some Key Questions* (Washington, D.C.: National Academies Press, 2001). Available online at books.nap.edu/html/climatechange/.
3. See the resources section in this book for citation information and URLs.
4. For both scientific and economic reasons, the Kyoto Protocol in fact regu-

lates a "basket" of key greenhouse gases. The methods for measuring and comparing them with CO_2 are complex and controversial, but they do not affect our basic argument here. For further information, see the reports of the IPCC.

5. Aerosols are airborne particles of various types, some of which (like sulfate aerosols) increase the reflection of sunlight and thus cause cooling (negative radiative forcing), and some of which (like soot from combustion) absorb heat as CO_2 does and thus cause heating (positive radiative forcing).

6. The specific figure here is from the Norwegian Polar Institute. "Polar Bears Facing Extinction," *The Independent* (London), 14 May 2002.

7. Integrated assessment models combine economic, atmospheric, and ecological models to allow quantitative projections of indicators such as emissions and average temperature, and to evaluate policies to mitigate climate change.

8. The TAR Working Group 2 SPM is at www.ipcc.ch/pub/wg2SPMfinal.pdf.

9. "The Carbon Logic" is posted at www.greenpeace.org/~climate/science/reports/fossil.pdf.

10. For a discussion of the issues here, see Stephen Schneider, "Can We Estimate the Likelihood of Climatic Changes at 2100?"*Climatic Change* 52 (2002): 441–51.

11. For a discussion of the likely probability distribution across the range of possible climate sensitivity values, see N. G. Andronova and M. E.. Schlesinger, "Objective Estimation of the Probability Density Function for Climate Sensitivity," *Journal of Geophysical. Research* 106 (2001):22,605–22; see also R. Knutti, T. F. Stocker, F. Joos, and G. Plattner, "Constraints on Radiative Forcing and Future Climate Change from Observations and Climate Model Ensembles," *Nature* 416 (2002): 719–23.

CHAPTER 3: FROM TEMPERATURE TARGETS TO EMISSIONS BUDGETS

1. The carbon cycle is the set of biological and geochemical processes by which carbon moves between living organisms, the soil, the atmosphere, and the oceans. The model used here is the Bern carbon cycle model, and the graph is from figure 25 in the Third Assessment Report Working Group 1 Technical Summary, available at www.ipcc.ch/pub/wg1TARtechsum.pdf.

Tom Athanasiou and Paul Baer

2. Among the critical uncertainties in the carbon cycle are the response of plants to increased atmospheric CO_2, the rate of increase of decomposition of plant matter with increasing temperature and changes in precipitation, and changes in biological and chemical absorption of CO_2 in the ocean. Again, consult the IPCC's reports for details.

3. For much more on "environmental space," a notion developed and popularized by Friends of the Earth International, see Michael Carley & Philippe Spapens, *Sharing the World: Sustainable Living & Global Equity in the 21st Century* (London: Earthscan Press, 1998); see also W. Sachs et al., *Greening the North: A Post-Industrial Blueprint for Ecology and Equity* (London: Zed, 1998). It's important to realize that even though the concept is based on science, "environmental space" is an extremely political notion; it represents the limit on pollution or use of a resource that we collectively (and, we hope, democratically) decide to accept. In the case of "atmospheric space," this decision will necessarily be a tough one, because there will be serious harm to humans and other species long before that "space" is "full."

4. As should the history of prediction in general. See, for example, George Orwell on "trend chasing;" in "James Burnham and the Managerial Revolution," *Collected Essays, Journalism, and Letters*, vol. 4 (New York: Harcourt Brace Jovanovich, 1968), 172–73.

5. "SRES" is short for "Special Report on Emission Scenarios." The SRES Summary for Policymakers is at www.grida.no/climate/ipcc/spmpdf/sres-e.pdf.

6. This graphic is from the Dutch National Institute for Public Health and the Environment's (RIVM's) "Keeping Our Options Open," which is available at www.rivm.nl/ieweb/ieweb/Reports/coolpolicybrief.pdf.

7. This insistence can be found in the Third Assessment Report itself, as well as in the SRES report.

CHAPTER 4: JUSTICE AND DEVELOPMENT

1. Or, alternatively, a fragile world. See Wolfgang Sachs et al., *The Jo'burg Memo: Fairness in a Fragile World, Memorandum for the World Summit on Sustainable Development.* (Heinrich Böll Foundation, 2002), available at www.worldsummit2002.de/downloads/memoE.pdf.

2. For a more philosophical approach to the argument for equal per capita rights,

see Paul Baer, "Equity, Greenhouse Gas Emissions and Global Common Resources," in *Climate Change Policy: A Survey,* eds. S. H. Schneider, A. Rosencranz, and J. Niles (Washington, D.C.: Island Press, 2002).

3. The A1 and B1 scenario families show population peaking at 8.7 billion in 2050, and then declining back to 7 billion by 2100, which would be an unprecedented drop in the absence of war, famine, or pestilence. By contrast, the most recent "medium variant" projections from the United Nations show the population at 9.3 billion in 2050 and still growing.

4. See, for example, *Living Planet Report 2002,* a new report from WWF. The report, downloadable at www.panda.org, found that exploitation of the Earth's renewable resources has grown by 80 percent in the past forty years and is now 20 percent higher than the natural capacity of the planet to replenish itself. But if current trends continue, it's only until 2050 before a second Earth would be necessary to meet human demand. It's not going to happen, of course, which is why the authors fear that global living standards will begin to plummet long before then.

5. Energy intensity is sometimes defined as the amount of energy required to produce a unit of GDP. In fact, a lot of energy isn't used in "production" but in final consumption, such as in heating, cooling, or transportation.

6. The title was taken long ago by *The Global Greenhouse Regime: Who Pays?* eds. Peter Hayes and Kirk Smith (New York: UN University Press, 1993).

7. Figures from *Definitions of "Equal Entitlements,"* Centre for Science and Environment, 1998. Based on UNFCCC, *Implementation of the Berlin Mandate, Additional Proposals from Parties,* 1997 (see www.cseindia.org/html/cmp/pdf/fact5.pdf), and, ultimately, the Brazilian proposal, which was tabled just before Kyoto.

CHAPTER 5: A PER CAPITA CLIMATE ACCORD

1. We focus on CO_2 because its long atmospheric lifetime, and indeed the long lifetime of energy infrastructure, makes it most urgent. The global treaty must also regulate other greenhouse gases, particularly methane, which has a high radiative forcing compared to CO_2. However, the non-CO_2 gases can't be treated in the same ways as CO_2, for many reasons.

2. The history of the idea, including the political context within which *Global Warming in an Unequal World* was published, can be found in *Green Politics* (New Delhi, India: Centre for Science and the

Tom Athanasiou and Paul Baer

Environment, 1999), the first volume of the CSE's Global Environmental Negotiations series. The second volume, *Poles Apart*, published in the spring of 2002, updates the analysis. See www.cseindia.org.

3. For Aubrey's history, see his *Contraction & Convergence: The Global Solution to Climate Change*, and also the extensive documentation available at www.gci.org.uk/.

4. *Green Politics*, 108.

5. *Rio to Johannesburg: Towards Concrete Action* (New Delhi, India: Tata Energy Research Institute, 2002).

6. The key term in all this is "well designed." For a recent review of the issues here, and details on how the Bush administration is trying to purge the U.S. sulfur-dioxide trading system of its regulatory backstops, in a manner that will lead to more "hot spot" problems for local communities, see Daniel Altman, "Just How Far Can Trading of Emissions Be Extended?" *New York Times*, 31 May 2002, C1.

7. Marcel M. Berk and Michel G. J. den Elzen, "Options for Differentiation of Future Commitments in Climate Policy: How to Realize Timely Participation to Meet Stringent Climate Goals?" *Climate Policy* 1 (2001):465–80. Available at www.gci.org.uk/papers/berkelz.pdf.

8. J.P. Weyant and J. H. Hill, "Introduction and Overview," *The Energy Journal* Special Kyoto Issue (1999): vii–xliv.

9. See, for example, F. Krause, P. Baer, and S. DeCanio, *Cutting Carbon Emissions at a Profit: Opportunities for the U.S.* (El Cerrito, Calif.: International Project for Sustainable Energy Paths, 2001), available at www.ipsep.org.

10. Christian Azar and Stephen H. Schneider, "Are the economic costs of stabilising the atmosphere prohibitive?" *Ecological Economics* 42 (2002): 73-80.

CHAPTER 6: TRADING, TAXES, AND FUNDS

1. See www.risingtide.org.uk/pages/carbon_action.html at the UK Rising Tide site.

2. See Gelbspan's "Toward a Real Kyoto Protocol," available at www.heatisonline.org. And note that some estimates of the expected proceeds from such a tax come in a whole lot lower than $300 billion. See, for example, the exchange on the Tobin tax in issue 133 of ATTAC's *Sand in the Wheels*, available at www.attac.org/attacinfoen/attacnews133.pdf.

3. See, for example, Romilly Greenhill, "Campaign for a Tobin Tax Gains Momentum," available at www.jubilee2000uk.org/finance/tobin-tax190302.htm.

4. World Council of Churches, "The Atmosphere as Global Commons: Responsible Caring and Equitable Sharing," November 2000, available at www.wcc-coe.org/wcc/what/jpc/cop6-e.html.

5. This is from the summary, which is available at www.ifg.org. The final text is forthcoming.

6. See the Website of the Just Transition Alliance, at www.jtalliance.org, for more information.

7. See www.skyowners.org for more information.

8. See www.sustainableeconomy.org/press/cleanenergy.htm for more information.

9 See www.doc.mmu.ac.uk/aric/aviation/DTQ_working%20paper.pdf for more information.

CHAPTER 7: THE FUTURE OF THE CLIMATE-PROTECTION COALITION

1. David E. Rosenbaum, "As Two Sides Push, Arctic Oil Plan Seems Doomed," *New York Times*, 18 April 2002.

2. See pages 90–122 of *Green Politics*, and our discussion of income and emissions in chapter 4.

3. Agenda 21 is on the Web at www.un.org/esa/sustdev/agenda21text.htm.

4. Anju Sharma and Apurva Narain, "Puppets on Purse Strings," *Down to Earth*, vol. 10, no. 23 (3 April 2002). Available at www.cseindia.org/html/dte/dte20020430/dte_analy3.htm.

CHAPTER 8: GLOBALIZATION AND GLOBAL WARMING

1. See the GATSwatch site for more than you ever wanted to know about the General Agreement on Trade in Services, and all that it portends, at www.gatswatch.org/GATSbasics.html.

2. Zhongxiang Zhang and Lucas Assuncao, *Domestic Climate Policies and the WTO*, available at papers.ssrn.com/paper.taf?abstract_id=288273.

3. Thalif Deenm, *Remove All Trade Barriers, Annan Tells Rich Nations, Third World Network*, 30 January 2001. *At* www.twnside.org.sg/title/barriers.htm.

CHAPTER 9: THREE FUTURES

1. For more on the road forward, take a look at *A Better World Is Possible: Alternatives to Economic Globalization*, a summary report from the International Forum on Globalization. You can download it at www.ifg.org.
2. For much more on the World Bank's funding of fossil energy projects, and the imperative of reforming the international financial institutions, see the Web site of the Sustainable Energy and Economy Network at www.seen.org.
3. You can download a copy of the Global Scenario Group's latest report, *Great Transition: The Promise and Lure of the Time Ahead*, at www.gsg.org.

INDEX

Tom Athanasiou and Paul Baer

ABOUT THE AUTHORS

TOM ATHANASIOU is a longtime green activist and technology critic, and the author of dozens of essays on environmental and techno-scientific politics. In 1996, his first book was published—in the United States as *Divided Planet: The Ecology of Rich and Poor*, and in England as *Slow Reckoning: The Ecology of a Divided Planet*. His interests focus on class division and distributive justice within finite environmental spaces.

PAUL BAER is a Ph.D. candidate in the Energy and Resources Group at the University of California, Berkeley. His research in the area of ecological economics focuses on both ecological and economic modeling and on the equity implications of various climate policy alternatives.

Athanasiou and Baer are the cofounders and coordinators of EcoEquity, an organization that aims to catalyze an honest public debate about the real demands of global environmental justice, and, more specifically, to campaign—within the United States—for a phased transition to a fair, global, second-generation climate treaty based on equal per capita rights to the atmospheric commons. EcoEquity publishes *Climate Equity Observer*, a free online newsletter available at www.ecoequity.org or by e-mail.